T0320737

THE EVALUATION GAME

Scientific research is communicated, organized, financed, governed, and evaluated through the process of publication. The result of this process is a highly competitive academic environment that rewards researchers for high-volume publication, preferably in high-impact journals, leading to the popularized expression "publish or perish." Universities and other scientific institutions are under similar pressure, with their aggregated research output being under constant scrutiny. This innovative text provides a detailed introduction to the origin and development of the scholarly metrics used to measure academic productivity and the effect they have upon the quality and diversity of scientific research. With its careful attention to both the positive and negative outcomes of research evaluation and their distinct expressions around the globe, *The Evaluation Game* guides the way to a more grounded understanding of metrics and the diverse academic cultures they give rise to.

EMANUEL KULCZYCKI is Associate Professor at Adam Mickiewicz University, Poznań, and Head of the Scholarly Communication Research Group. From 2018 to 2020, he was the chair of the European Network for Research Evaluation in the Social Sciences and the Humanities, and in 2019, he co-founded the Helsinki Initiative on Multilingualism in Scholarly Communication. He has been a policy advisor for the Ministry of Science and Higher Education in Poland since 2013. He co-edited the *Handbook on Research Assessment in the Social Sciences* (2022).

THE EVALUATION GAME

How Publication Metrics Shape Scholarly Communication

EMANUEL KULCZYCKI

Adam Mickiewicz University

Shaftesbury Road, Cambridge CB2 8EA, United Kingdom

One Liberty Plaza, 20th Floor, New York, NY 10006, USA

477 Williamstown Road, Port Melbourne, VIC 3207, Australia

314–321, 3rd Floor, Plot 3, Splendor Forum, Jasola District Centre, New Delhi – 110025, India

103 Penang Road, #05–06/07, Visioncrest Commercial, Singapore 238467

Cambridge University Press is part of Cambridge University Press & Assessment, a department of the University of Cambridge.

We share the University's mission to contribute to society through the pursuit of education, learning and research at the highest international levels of excellence.

www.cambridge.org
Information on this title: www.cambridge.org/9781009351188

DOI: 10.1017/9781009351218

First published 2023

A catalogue record for this publication is available from the British Library.

ISBN 978-1-009-35118-8 Hardback
ISBN 978-1-009-35119-5 Paperback

Contents

Acknowledgments

Writing this book was a luxury. As I have shown in the pages of this story, in academia, there is often no time for research work and very little time for thinking. Thanks to a five-year grant from the National Science Center in Poland (UMO-2017/26/E/HS2/00019), I was able to carve out time not only for research but especially for thinking about evaluation and the future of academia. I am aware of what a rare privilege this is, and I am grateful for it.

While working on this book, I benefited from the help of many people who supported and inspired me. Not wanting in any way to create a ranking of those to whom I owe a great deal, I will list them in alphabetical order.

Przemysław Korytkowski and Krystian Szadkowski had the greatest impact on my thinking about science as a research subject and about the influences exerted by science policy, academia, and research evaluation. They have been my ideal discussants. In every conversation with Przemysław, I have to make sure that I have a concrete fact or figure, or other form of empirical evidence to back up each of my statements. Debates with him are extremely inspiring and challenging. Krystian forced me to reject fixed views and inspired me to look at writing this book as primarily a conversation with myself that would allow me to rethink my position in order only then to communicate my findings with readers. He was also the first reader of the entire manuscript, and on his advice, I turned parts of the book upside down.

I could not have completed this book if I had not had the opportunity to work with great scholars. Members of the Scholarly Communication Research Group at the Adam Mickiewicz University in Poznań were early commentators on the first drafts of chapters. More importantly, for several years I had the chance to discuss my ideas at daily shared breakfasts and lunches. I would like to thank Marek Hołowiecki, Franciszek Krawczyk, Jakub Krzeski (who with Anna Piekarska prepared the book's index), Ewa A. Rozkosz, Aleksandra Swatek, the already mentioned Krystian Szadkowski, and Zehra Taşkın.

I would not have been able to develop my understanding of research evaluation systems and the situation in different countries without my friends and colleagues from the European Network for Research Evaluation in the Social Sciences and the Humanities, which I had the honor of leading thanks to the trust and support of Ioana Galleron and Geoffrey Williams. Working together for four years has allowed me not only to further my understanding of how evaluation technologies work in different countries but also to gain an inside view of how these technologies are designed.

My thinking about research evaluation in diverse contexts and details of how particular national systems work has been influenced by help and conversations with the late Paul Benneworth, Andrea Bonaccorsi, Tim C. E. Engels, Elea Giménez-Toledo, Raf Guns, Gaby Haddow, Jon Holm, Thed van Leeuwen, Andrey Lovakov, Michael Ochsner, Ginevra Peruginelli, Janne Pölönen, Vidar Røeggen, Gunnar Sivertsen, Lin Zhang, and Alesia A. Zuccala.

Thinking and writing in a foreign language are quite challenging. I would like to thank Saskia Fischer, who has been essential in the preparation of this book, not only with her editorial and linguistic support but also for sensitizing me to the many cultural dimensions of expressing my ideas in English. In this place, I would like to thank Mikołaj Domaradzki who taught me how to write academic papers in English and aim high.

Finally, I would like to thank my family, without whose support not only would I never have been able to complete this book but without whom I simply would not have wanted to write it. My wife Marta and sons Ryszard and Tadeusz give me joy in the work I do. Their emotional support is the most solid foundation for me. To the three of them, I dedicate this book.

Introduction

Weigl used to do everything perfectly, whatever it was. He spent money on only two passions: fishing and archery. However, the fishing rods, trout flies, and bows and arrows that he acquired on his travels around the world never quite met his standards. For him, the only way to be sure of having good equipment was by designing it oneself. In his research, he was just as much of a perfectionist as in his passions, and he mobilized his considerable ingenuity to do whatever was necessary in order to achieve his goals. At the beginning of the twentieth century, researchers were trying to force lice to eat germs in order to create a vaccine against typhus – a deadly epidemic that killed more of Napoleon's soldiers as they retreated from Moscow in 1812 than the Russians did. Weigl, with characteristic inventiveness, literally turned their reasoning upside down, saying "we'll stick them up its [the louse's] ass" (Allen, 2015, p. 22).

Rudolf Weigl – a Polish biologist from Lviv – invented the first effective vaccine against typhus, which saved countless lives during the interwar period. As part of his research, he employed people to serve as lice feeders; they had cages with typhus-carrying lice strapped onto their thighs. These feeders were most often Polish intellectuals, Jews, and underground fighters. In the group of over 2,000 feeders, there were two outstanding men: a mathematician, Stefan Banach, the founder of modern functional analysis, and a microbiologist, Ludwik Fleck, who developed the concept of "thought collective" crucial for Thomas Kuhn's later notion of the "paradigm shift." Weigl was nominated many times for the Nobel Prize, but he was never awarded it. Once, he refused to be a candidate because according to him, his discovery was not among the highest-ranking ones. Another time, during World War II, he opposed becoming a candidate from Hitler's Germany (Wincewicz et al., 2007). For Weigl, it was the discovery that was science's most important outcome, not awards or publications.

Weigl's approach to publishing academic papers was radical. He believed that the actual research consisted in doing science and making discoveries, whereas writing was torture and a waste of time. While a student, he had been forced by his supervisors to publish papers just to keep receiving his scholarship. However, he drove his own students to despair because his way of doing science reduced the number of publications they could put on their résumés (Allen, 2015). What is now called "salami publishing" in science – that is, dividing a large study that could be published as a single research paper into smaller papers – Weigl called "duck shit": just as ducks leave a lot of traces while walking about in the yard, scientists hastily publish articles with partial results that are the product of undeveloped thought. It was Weigl's belief that the true value of research manifested itself in its impact on society and not in publications. Being neither the means of exerting an influence on society, publications were also not one of science's key pillars. And yet today, in most academic environments, they are considered the engines of science's development.

Publication-Oriented Science

Today's science is publication oriented; it is communicated, organized, financed, governed, and evaluated through publications. Seen from the current centrality of publication to science, Weigl's approach appears idealistic and unconvincing. I would put it even more forcefully: Many researchers and policy makers perceive science as confined to publications, forgetting the real people who do the research, make discoveries and inventions. These are the people who work in institutions that together constitute "academia." However, academia should not be understood as a kind of collection of geniuses but rather as an international collective endeavor that involves thousands of researchers, technicians, students, and administrative staff members working in a wide variety of places.

Nonetheless, the image of science as the work of geniuses is quite popular, and the lives and views of well-known scientists are used to promote differing visions as to how science should be organized, financed, and governed. In 2013, Peter Higgs – a Nobel Prize laureate in physics – declared that no university would have employed him in today's academia because he would not be considered "productive" enough (Aitkenhead, 2013). This statement has been used as evidence that a "publish or perish" research culture has conquered science, and to back up the argument that academia must cease to be publication-oriented. Under a culture of publish or perish, academics are pressured into producing large numbers of publications, not only to succeed but merely to survive in their work environments. Yet academia has been publication oriented for many years. In her tribute to William J. McGuire, social psychologist, Banaji (1998) refers to McGuire's bet with a colleague that he would not publish a single paper until after receiving tenure at the University of Illinois. McGuire was tenured in 1960 and, one year later, published ten papers that he had already written but had not submitted for publication.

Banaji (1998) mentions that "this act of daring made him an instant hero of many of us when we were graduate students."

Although such arguments are compelling, they rarely offer solutions for how researchers can survive in academia. Instead, they end up reducing the discussion to the level of the absurd. Taking such arguments even further, one could say that Albert Einstein would not have been awarded tenure in the present days given that his major works were not published in English or – and here we reach the ultimate phase of pure absurdity – that Socrates would not have been granted a project because he did not publish any papers. In this way, the academic community uses ahistorical reflection to build various myths which it then deploys in its defense of the profession's autonomy.

Since the Manhattan Project started to produce the first nuclear weapons during World War II, science has changed irrevocably. In the mid-twentieth century, it entered fully into the era of big science which was characterized by a rapid growth in the number of institutions, researchers, discoveries, and publications (Price, 1963). At the same time, discussions on the role of science in society started to be shaped by definitions of scientists' responsibilities toward society. As Shapin (2012) shows, there never was an Ivory Tower and yet the call for scientists to leave it became the leitmotiv of twentieth-century reflection on the impact of research.

Research initiatives like the Large Hadron Collider or the Gran Telescopio Canarias need financial support from many countries. Because of this, science has begun to be both financed and carried out internationally. Science's ongoing development requires tools through which governments can distribute public funds and evaluate the results of provided inputs. However, science, the research process, and scholarly communication are too complex to have all their features reduced to a single model. Because bureaucrats seek clear and undemanding solutions with which to justify their decisions, the models elaborated for this purpose need to be simple and based on readily comprehensible elements. They should, moreover, be easy to explain to the general public. It is thus very tempting to use publications as science's touchstone: one can easily count them and say who has published more.

Scholarly communication is one of the key pillars of science. However, it does not only manifest itself in the publication and dissemination of research results through academic papers. Scholarly communication begins with reading, discussing, and arguing. Presenting and reviewing preliminary research results are also important phases of scholarly communication which does not end at the time of publishing. Indeed, as a circular process, scholarly communication cannot be reduced to any one of its phases. Yet in actuality, it is subjected to precisely such a reduction, which in fact takes place twice. First, when researchers and policy makers start to think that disseminating research results means only publishing academic papers. Second, when they identify journal articles as the key elements of research and science themselves, to which they can be reduced. Finally, researchers and policy makers end with publications that are abstracted from research which represent the science. This synecdoche represents a pivotal feature of today's academia. For instance, Dahler-Larsen (2015) argues that one of evaluation's constitutive effects is to provide an erroneous image of what is actually going on, and to support the overarching assumption that research should be understood as production (e.g. of new papers, patents etc.). More importantly, such abstracted publications can be counted and the results can be used to justify various decisions. Nonetheless, someone still needs to decide how publications will be measured and what measures will be applied. In this moment of deciding what (e.g., all academic publications or only peer-reviewed ones?), by whom? (e.g., policy makers, researchers), and how (e.g., should all co-authors take full credit or should it be distributed among authors?) publications are evaluated, the power of measures is established. Measures and measuring are the technologies of power (i.e., instruments embodied as sets of protocols, indicators, and policy aims) that are used by global actors, states, and research institutions to evaluate science and through this, to fund, control, and govern the whole science sector.

For Weigl, counting publications in order to measure the true value of research constituted an offense against of the very essence of science itself. Today, the "metricization" of science, that is the introduction of metrics into research and academia, is global. A great deal has been written about this phenomenon: starting from studies on the quantification of social practices (Espeland & Stevens, 2008; Muller, 2018), histories of measuring science (Godin, 2005, 2009), audit cultures in higher education (Power, 1999; Shore & Wright, 2003; Strathern, 1997), research on the impact of using indicators in science (De Rijcke et al., 2016; Müller & De Rijcke, 2017), monetary reward systems in academia (Quan et al., 2017), through critiques of university rankings (Pusser & Marginson, 2013; Sauder & Espeland, 2009; Yudkevich et al., 2016), to the consequences of local uses of research evaluation systems in various countries (Aagaard, 2015; Aagaard & Schneider, 2017; Butler, 2003b; Kulczycki, Rozkosz, & Drabek, 2019). The community of scholars that focuses on using metrics for monitoring, reporting, managing, and – most often – evaluating research and researchers, produces a discourse that shapes science policy and has an impact on research and academia. Thus the way in which metricization based in publication metrics is discussed influences the system of science itself.

Academia constitutes a complex system which has its own history and heritage. Although science is global, inside this global system, various local, institutional, national or regional structures can be found, and each of these has its own specific background and features. Discussions on using measures, metrics, and evaluating research show what tensions might occur when local or national conditions meet global values and challenges. Moreover, these discussions reveal how the demands of the global market in higher education can change researchers' work and ways of communicating their research. Nonetheless, the responses – or in other words, implementations of global demands – vary by virtue of their dependence on specific cultural and historical contexts.

Aims of This Book

In this book, I aim to offer an alternative position to the discourse on using publication metrics and measuring science that is being produced by the community of research evaluation scholars. This book focuses on research evaluation systems to make scholars and policy makers aware of two key blind spots in this discourse.

The first blind spot relates to the absence of the Soviet Union and post-socialist countries in the histories of measuring science and evaluating research, despite the fact that these countries have played a key part in this history from its very inception. In these countries, a distinct discourse on using metrics in the system of science – based on the scientific organization of scientific labor and central planning – was articulated.

The second blind spot relates to thinking about global differences in studies of the transformations of scholarly communication. I show that the contexts in which countries face the challenges of publish or perish culture and questionable (or so-called predatory) journals and conferences, should be taken into account in discussions about them. In order to identify and explore both blind spots, we must therefore include state(s) as an explanatory factor in the study of the effects of research evaluation.

Moreover, I argue that in today's academia, research is influenced not only by the global context of the knowledge-based economy or the idea of accountability in public funds. The way in which research is done can also be mediated through national and local science policies. For instance, in some countries, among them Canada, France and the United States, researchers experience research evaluation mostly when universities measure their publication productivity or when they are faced with a tenure procedure. In other countries, for instance in Australia, Argentina, Norway, Poland and the United Kingdom, research evaluation is also experienced through national research evaluation systems which influence both university strategies and tenure (or tenure-like) procedures.

These blind spots inspired me to elaborate an alternative perspective on the history of research evaluation and its effects going forward. This perspective is based on the concept of the "evaluation game," which refers to the ways in which researchers, academic managers, and policy makers react on being evaluated and when they act as evaluation designers. While, as stated above, the research evaluation game is a global phenomenon, it was not my intention to write a book about the global phenomenon of research evaluation and measurement that would attempt to account for every distinct situation in every part of the globe. The world of science is far too complex to capture in a single book. Rather, as a member of the community of research evaluation scholars, my aim is to intercede within the discourse on research evaluation in a way that promotes a deeper understanding of the social world of science, thereby contributing to creating a better environment for research and researchers.

In this book, I conceptualize doing research and managing academia along three intersecting planes, in each of which the simultaneous significance for doing science is also emphasized: (1) global (supranational), (2) national, and (3) institutional/local. In this way, this is a *glonacal* perspective (see Marginson & Rhoades, 2002) in which conceptualized phenomena are characterized by global, national, and local planes.

The first plane involves global transformations of academic labor (e.g., the pressure to produce more and more publications or to publish mostly in English). The main actors here are supranational institutions like the World Bank, the Organization for Economic Co-operation and Development (OECD), global companies producing bibliographic and bibliometric data sources like Elsevier and Clarivate Analytics, and institutions producing university rankings such as Shanghai Ranking Consultancy (Shanghai Ranking) or Times Higher Education (World University Rankings). These organizations transcend the state form and through their own global influence play an important role in the universalization of publication metrics in science as well as in shaping national systems by providing global exemplars and solutions like field classifications or bibliometric indicators (cf. Godin, 2005). In this book, I argue that one can observe some of the manifestations of characteristics specific to countries of the Eastern Bloc or to Western countries at this supranational level.

On the second plane, research evaluation systems influence the day-to-day work of members of the academic community in a given country. At the same time, through their administrators and policy officers, states have to react to and interact with transformations and values manifesting at the global level that might cause varied tensions. These might, for instance, include tensions around the promotion of publishing research results in English (a goal of scholarly communication rooted in global-level values) or national languages (a goal rooted in state values that cultivate local culture and heritage or aim to reduce inequalities in academia). In this book, states (especially at the national levels) are perceived as characterized by a high degree of sovereignty and agency. This view might be criticized as ahistorical or even essentialist (cf. Jessop, 2002) considering that the state has lost its sovereignty and has become more of an instrument in the hands of various – often global – interest groups. In other words, one might say that globalization has weakened the agency of the state and made it rather a medium of and for the global context. Moreover, New Public Management – the group of ideas that is transforming relations across government, public institutions, and society – might be perceived as a departure from the traditionally understood state and its central (not in the Soviet sense but in the sense of the welfare state) management of the public sector.

On the one hand, I second this view of the state's weak position. In Chapter 2, I argue that in fact the state should not be the basic unit of analysis and could instead be replaced by a broader unit, that is the world-system (Wallerstein, 2004). On the other hand, however, the state functions as an actor on the global stage that is both influenced by the global context and influences its own local contexts. This is why I argue that doing research and managing academia in the era of research evaluation systems should be understood as they are in this book: taking place through three intersecting planes, that is the supranational, national, institutional and locals (levels of practice). At the third level of academic practices, academia's institutions and people work, react, and adapt to changing conditions of academic labor. In this book, I am especially interested in reactions and resistance (at the local level) caused by state policy instruments, that is by research evaluation systems (at the national plane) influenced by global transformations (at the global plane).

Building on the above elaboration of the three intersecting planes, this book proposes a new theoretical framework for understanding the effects of research evaluation systems and using metrics in academia. I have written it in order to explore how the process of evaluating science – mostly through the prism of publications – shapes the production and communication of scientific knowledge in and through universities and research institutes, in a way that renders them akin to political institutions of the state. In asserting this, I am referring to the fact that they require state resources, the state provides them with certain benefits, and institutions in turn gain some authority from the state (Neave, 2012).

Research evaluation systems are science policy instruments used to measure academia's performance. I use the term "academia" to indicate members of a community who share common beliefs, norms, and values regarding science and research. This community includes primarily (1) professors and researchers from higher education institutions and research institutes as well as (2) these institutions and (3) their staff management. Individuals became a part of academia when they learn how to act (and what to believe) in a way that is acceptable to the academic community. Institutions become part of academia when they are defined as political institutions of the state in the higher education or science sectors.

Turning now to the effects produced by the operation of research evaluation systems, these can be either intended and unintended. Intended effects can be understood as goals accomplished and successful public interventions. However, when it comes to investigating the unintended effects, this cannot be reduced to tracking and reporting the unforeseen or unpredicted side effects of preplanned interventions. This is due to the fact that unintended effects originate not only in social interventions themselves but also – among others – in the context in which such policies are implemented (e.g., unstable conditions of academic labor or scholarly communication reduced to publishing papers). Moreover, the intentions of policy designers, policy makers, or stakeholders are rarely explicitly communicated. Thus it can be difficult to assess whether certain effects were intended or not. Within evaluation studies, the distinction between intended and unintended has been criticized (Dahler-Larsen, 2014) from the position that all effects exert an impact on the evaluated reality. Nonetheless, I argue that this distinction might still be useful for understanding the science policy perspective through which policy makers assess the results of their efforts.

There is a long tradition, within both administrative and organization theory, of exploring the dysfunctional consequences of performance measurement (Ridgway, 1956). During the first half of the twentieth century, the byproducts and impact of performance measurements were analyzed in numerous areas that ranged from American and Soviet industries to public policies. Even then, studies showed that the use of a single measure was not adequate and should be replaced by the use of composites, that is multiple and weighted criteria or – as was later suggested – by multiple indicators. This was because no single indicator could ever reveal more than a small part of the multidimensional picture that is composed through research (Martin, 1996).

This knowledge and experience were utilized within New Public Management (Hood, 1991), which transformed performance measurement substantially into outcomes-based performance management (Lowe & Wilson, 2017). For researchers on evaluation systems, New Public Management is an often-cited reference point, which designates the central implementation of research evaluation systems. Its mention serves to emphasize a process of transforming relations across the government, public institutions, and society. In the second half of the twentieth century, these relations shifted, moving from an era of professional autonomy coupled with bureaucratic systems toward the promotion of efficiency in the production of public services (Lowe & Wilson, 2017).

I use the theoretical framework that I elaborate through three key steps to explore transformations of scholarly communication caused by the process of measuring and evaluating science.

First, I present the concept of the *evaluative power of the state* as a ground for developing the framework in which the effects of research evaluation systems can be investigated. Evaluative power is the capacity of the state to influence and shape the key area and to change practices of individuals and institutions. While evaluative power can also be identified as the power of global actors (e.g., companies developing citation indexes like Web of Science Core Collection [WoS] or Scopus), in this book, I focus mostly on evaluative power as a characteristic of states.

Second, I present the concept of the *evaluation game* through which the reactions provoked by evaluative power manifest themselves. In this way, the evaluation game is one of academia's responses to the changing context of academic labor. For instance, some forms of evaluation game manifest through the establishment of new and questionable journals or publishing within them only to fulfill expectations produced by the research evaluation regime. Thus I demonstrate that the evaluation game and its consequences are (un)intended effects of the design and use of research evaluation systems.

Third, I rethink the history of the measurement and evaluation of science and argue that understanding the consequences of research evaluation systems better requires the incorporation of an omitted part of this history. In other words, I show that performance measurement in the science sector is not only a hallmark of Western science but was in fact first implemented at a national level in Russia, and then later in the Soviet Union and Eastern Bloc countries. Bringing this heritage into the spotlight is a necessary step for understanding why in countries with similar research evaluation systems – like Australia and Poland, for example – the reaction and resistance against the systems manifests itself in diverse forms of evaluation game, and further, why researchers perceive the same elements of those systems (like the use of bibliometric indicators or peer review) in substantially different ways. For instance, as Mishler and Rose (1997) show, distrust is the predictable legacy of Communist rule and in the postcommunist societies of Eastern and Central Europe, trust in experts is substantially lower than in other societies. Thus, when in the 1990s, Poland implemented sweeping reforms in the science sector, skepticism about peer review was one of the key obstacles to rolling out a performance-based research funding system (cf. Jablecka, 1997). Further, Sokolov (2020, 2021) argues that in Russia, the use of quantitative indicators was an expression of distrust by the state of scientists' capacities to act as evaluators. Moreover, the study of research evaluation history in Russia and Eastern Bloc countries is made more relevant by the fact that the current wave of Chinese modernization is marked by Soviet heritage.

Finally, building on the above three discussions, I go on to examine how research evaluation systems shape scholarly communication in contemporary academia, and how various practices evident in the evaluation game can be used as tools for understanding these changes.

Power and Its Dark Side

This book is the result of research carried out at the intersection of three perspectives. When I started to investigate research evaluation systems, I was a philosopher interested in the communication mechanisms and policy agendas that shape academia in the context in which I worked. At a certain point, I became a social scientist who moved from philosophy to social sciences and investigated scholarly communication, science policy, and research evaluation. In this way, I combined two perspectives. First, the perspective of a researcher who works in academia and whose work is evaluated and performance measured. Second, the perspective of a social scientist who critically investigates research evaluation systems and scholarly communication as they are implemented through various bibliometric and scientometric indicators. In consequence, I was being measured at the same time as I was also measuring my peers, for instance, by showing how publication patterns in social sciences and humanities were changed by research evaluation systems (Kulczycki et al., 2018, 2020). Thus I investigated research evaluation systems and criticized some science policy instruments by arguing that the process of constructing the measures served as means of sustaining the evaluative power of the state. These two perspectives allowed me to understand my own situation in academia better.

At some point, however, my critique of the Polish research evaluation system was recognized and acknowledged by policy makers. Having conducted two research projects on the effects of research evaluation, I was asked to serve as a policy advisor and – for some period – as a policy maker. In other words, I was put in a situation in which I was able to use my critical studies and recommendations to suggest how the Polish research evaluation system could be improved. I decided to take on this task and help – in the eyes of some of my colleagues – "power's dark side" to reproduce an oppressive system which I had been examining and criticizing for years. This third policy perspective showed me how difficult it is to design, implement, and use various science policy tools and why each social intervention always triggers both resistance and other types of reaction.

The experience that I gained from being located at the intersection of three standpoints showed me first, that while a dialogue between the academic community and policy makers might be fruitful, it is a demanding task and second, that in their current form, policy instruments are strongly shaped by researchers' demands. Before joining this dark side of power, I had thought that all the poorly designed policy instruments were the product of policy makers' own work and that they never listened to researchers and did not care about their opinions. Later, I learned that policy makers often design solutions exactly as suggested by the academic community for which "my own field's perspective" is the only acceptable perspective. The problem is that some researchers are not aware of the structural effects produced by changing only one element of the system, and it is therefore difficult for them to formulate useful recommendations for policy makers. On the other side, policy makers do not always understand that the solutions suggested by natural scientists might not work in the humanities and they should therefore engage researchers from all the fields for which they design policy solutions. Nonetheless, the fact remains that as actors in the public sector, policy makers and academia need each other. Thus, improving discussions between them and promoting mutual understanding would be beneficial for all sides.

In this book, I put these experiences to use. As a researcher who could describe himself as a critical social scientist working across disciplines, I believe that the results and recommendations that we draw from our studies should be communicated and used to improve our social worlds. Revealing power relations should not be the ultimate goal, but rather one of the steps along the way. Thus if, as a product of their research, social scientists identify things that do not work, as well as the reasons for this, they should suggest ways of improving the situation. It is much easier merely to reveal power relations, rather than to reveal them and suggest ways of tackling them, all the while taking responsibility for our recommendations. And yet I believe that social studies, especially those critical investigations that reveal power relations, are extremely useful and relevant, because the first step always requires that we develop an understanding of the object of our critique. However, this step should then be followed by proposals for possible solutions for improving social interventions and making the use of technologies of power more responsible.

I now outline the book's central themes by posing and answering a series of key questions. In doing so, I underline the contribution that an exploration of the effects of research evaluation can make to our understanding of academia today.

What is the research subject of this book? I examine the evaluation game as the effect of research evaluation systems that are situated in the global context in which research is done. The evaluation game is a social practice that is transformed by an implementation of or change in research evaluation regimes in academia. As such, it is one of academia's forms of resistance or response (reaction) to the power of the research evaluation system that measures and evaluates science.

Why study research evaluation systems? Academia is today under intense pressure to be accountable and as a consequence, every aspect of academic work is quantified, monitored, and governed. Among other things, this pressure translates into exploitation within academia, which can provoke health problems and burnout for researchers. I point out how the evaluation of research and the use of metrics are indicated as some of the key reasons for which the situation in academia is as it is. Around the world, researchers are evaluated and assessed through varied procedures such as the assessment of manuscripts, grant proposal evaluation, or tenure track processes. However, in some countries, these procedures are complemented by all-encompassing national systems that are legitimized or produced by states. These systems have the power to transform researchers' daily practices to a much greater extent than other regimes that include the peer review of articles or grant proposals. The analysis I offer makes it clear that research evaluation systems have a substantial transformative power and that the transformations caused by them are determined by historical and cultural factors.

Are you for or against research evaluation based on metrics? This is one of the questions most frequently addressed to investigators in research evaluation studies. Many researchers see the issue in black or white terms: If you conduct scientometric analysis, it means that you are in favor of research evaluation and cannot be one of us, that is a "real researcher" for whom any form of quantitative evaluation is synonymous with oppression and damaging to the freedom of science. The question of whether one is for or against metrics is one of the central questions for me, and I pose it here because my answer to it provides a basis for understanding the approach I take in the study. So then, am I for or against metrics in evaluation? I am for metrics in research evaluation because – as I argue in this book – they are an unavoidable part of today's science system and furthermore, they have the potential to be useful for all those within it. However, the core reason for I am not against the metricization of science it that, as I stated earlier, it is not the problem. The problem is economization that promotes the idea that science's economic inputs and products should be utilized for boosting the economy. Metrics cannot be abandoned due to the fact that – as I argue in Chapter 2 – quantification is an inevitable element of the modern social world. However, researchers and policy makers can work on transforming the tendency to attribute economic value to all activities in academia.

What approach do you take? Science is a social practice and as such is reproduced by the actions that we take in line with the rules, norms, and values shared by the academic community. As social actors, we not only act and solve problems but also constitute our social worlds. This is why I start from the assumption that research evaluation is not a bipolar phenomenon in which those who evaluate come into contact with that which is evaluated. Rather, research evaluation is a constitutive process of producing evaluated objects in and through power relations, designing metrics, and transforming the context in which evaluation takes place.

What are your findings? When the state implements a research evaluation system, academia responds by establishing various types of evaluation game. The shape of such a game depends not only on the system itself but also on the historical and cultural context of its implementation. Contrary to dominant accounts, the first research evaluation system was not established in the United Kingdom in 1986, and later rolled out in other Western European countries and Australia. Instead, the first research evaluation system was in fact implemented over 50 years earlier in the Soviet Union (and to some extent even earlier, in the era of Imperial Russia) and in countries of the Eastern Bloc. This book shows that the socialist heritage plays an important role in shaping academia's response to today's evaluation regimes and the ways in which metrics are constructed for evaluation purposes. My research shows that playing the evaluation game is an inevitable aspect of any metrics-based research evaluation system and as such, will transform scholarly communication. Academia is obsessed with metrics because economization (mostly neoliberally oriented) drives the state to introduce instruments which cater, above all, to the values of productivity and accountability. If we (as members of the academic community) cannot avoid the evaluation game, we should still try to change its underlying logic of economization that drives today's academia. For instance, instead of being focused on a linear input–output model of economic development founded on values that center on individuals, competition among them, and profit margins, we could focus on the common (or public) good based on the values of cooperation, solidarity, and well-being. In such a new logic, some metrics would also be used which would transform social practices. More important, however, is the question of what values would drive academia in such transformations.

What is the significance of this study? The literature on research evaluation systems has focused almost exclusively on their effects in the area of publication productivity and researchers' approaches to evaluation regimes in Western countries (i.e., Australia, Denmark, Finland, Italy, Norway, and the United Kingdom). The frameworks used in these studies are based on the assumption that evaluation regimes transform research and publication practices because metrics themselves are reactive and change how people think and act through the internalization of evaluation rules. Moreover, the studies often assume that people within academia approach metrics and peer review (as two poles of research evaluation methods) in identical fashion, across the whole science system and regardless of geographical context. In this book, I construct a different framework that shows that evaluative power rapidly transforms daily practices in academia and by doing this, pushes people to adapt to a new context. This adaptation causes various kinds of reaction that are manifested in diverse types of evaluation game. The framework laid out in this book allows me to draw attention to the fact that similar technologies of evaluative power (i.e., the instruments and metrics of research evaluation systems) have different effects depending on the historical and cultural contexts of implementation.

What are the implications of the study? The implications of this study are twofold, that is, both epistemological and prescriptive. From an epistemological standpoint, this study shows that centrally planned science was a research evaluation system before New Public Management inspired the first such systems in Western countries. My findings underscore the fact that, from a science policy perspective, the important thing is not whether the evaluation game is intended or unintended. This is because the game develops as a response to evaluation regimes and transforms our social world of science. Moreover, the concept of the evaluation game shows that such a game is established through a dialectical process in which the state or another actor that holds power introduces rules and metrics. For their part, researchers devise various strategies for following these rules at the lowest possible cost to them. This is due to the fact that in a game, it is rational to put in as little effort as possible (e.g., the smallest number of moves) in order to achieve one's goal. From a normative standpoint, I want to show that designing better and more comprehensive metrics for research evaluation purposes is not enough to stop various questionable research practices like the establishment of predatory journals, guest authorship, or ostensible internationalization, often considered as "gaming" the research evaluation regimes. It is not metrics but the underlying economization that is the source of the transformation of scholarly communication and of academia itself. With this book, I want to show that a greater understanding of the reasons for which research practices are transformed may lead policy makers, stakeholders, and academics toward better solutions for governing academia and articulating the values that should guide management. This is a critical task today because the pressures upon academia are constantly increasing, because more and more countries are implementing or considering introducing evaluation regimes.

It is my hope that this book can lead us to a better understating of what role measurement and the evaluation of research play in science. There is no way to do publicly funded research and avoid (ex-post or ex-ante) evaluation. This being the case, it is crucial that we ask how we can impact science policy and create more responsible technologies of power.

Structure of the Book

Research evaluation systems are an important part of today's academia. The evaluation game is complex, dynamic, and rooted in various historic and cultural backgrounds. Thus the effects of measuring social processes cannot be fully understood through a single theory. This book's central thesis is that understanding the effects of research evaluation systems on scholarly communication requires that we identify how an evaluation process constitutes the objects it measures and how the context in which the evaluation is implemented determines the shape of this process.

The book owes its structure to the belief that we must answer the following critical questions if we are to understand how research evaluation shapes scholarly communication: Who has the power to produce evaluations? Which conditions and historical contexts allowed the rise of research evaluation systems? And finally, how did diverse evaluation systems produce different practices of resistance and adaptation in academia?

The book consists of six chapters, devoted to research evaluation systems and playing the evaluation game. Chapter 1 begins by introducing research evaluation as a manifestation of evaluative power. This chapter describes how evaluative power is legitimized and how it introduces one of its main technologies, that is research evaluation systems. A definition of "games" as top-down social practices is put forward and – on the basis of this framework – I present the evaluation game as a reaction or resistance against evaluative power.

Chapter 2 is dedicated to presenting the background and conditions that made the rise of research evaluation systems possible. Thus the faith in numbers, the construction of metrics and transformations in the very nature of academic labor are discussed. I show that economization and metricization underlie contemporary understandings of research performance and productivity. I argue that beyond the general phenomena that characterize modern society, such as rationalization, capitalism, or bureaucratization, there are also other things that could be presented as constitutive conditions for evaluative power. To this end, I provide a sketch of how today's academia is defined in relation to evaluation-related concepts such as audit cultures, the neoliberal university or the culture of publish or perish. These concepts – linked to New Public Management – can often be detected in the background in periods in which research evaluation systems are introduced.

Chapter 3 explores important background for understanding research evaluation systems in Central and Eastern Europe. This is the beginnings of the scientific organization of scientific labor and the development of scientometrics in the first half of the twentieth century. Further, this chapter shows how the history of research evaluation has been written mostly from a Western perspective that neglects science in the context of the Soviet Union and countries of the Eastern Bloc. The chapter therefore provides an in-depth analysis of research evaluation within the centrally planned science of the Soviet Union and countries of the Eastern Bloc. Research evaluation systems are often described technologies that came into existence 40 years ago as new ways of establishing relations between the state and the public sector. In this chapter, however, I demonstrate that centrally planned science introduced a national (ex-ante) research evaluation system and the assessment of research impacts decades before the rise of New Public Management and the first Western European systems.

Chapter 4 examines the diversity of evaluative powers and research evaluation systems along the three planes I earlier identified, that is the global, national, and local. This chapter starts with an explanation of why the Journal Impact Factor (JIF) has become the most popular proxy of the quality of research. Next, varied international citation indexes and university rankings as technologies of power are analyzed. In this chapter, I focus on representative national systems as implemented in Australia, China, Nordic countries (Norway, Denmark, Finland), Poland, Russia, and the United Kingdom. In addition, research evaluation systems designed for individual researchers are discussed. In the chapter, I examine not only the similarities and differences in the ways in which systems are designed and operate but also the ways in which they are criticized and the critiques responded to.

Chapters 5 and 6 deliver an analysis of the evaluation game as a response to evaluative power. In Chapter 5, I describe the game's main players, that is policy makers, institutions, managers, publishers, and researchers. Moreover, the chapter addresses the challenge of attributing causality to research evaluation systems and of distinguishing *gaming* from *playing the evaluation game*. Recognizing an activity as *gaming* or *playing the evaluation game* is not straightforward. The same activity (e.g. publishing in a predatory journal) may be considered *gaming* when it serves to maximize profits, or *playing the evaluation game* when it fulfills evaluation requirements, where the stakes in the game are not related to financial bonuses but to maintaining the status quo in redefined working conditions.

Chapter 6 chapter deals with the main areas in which the evaluation game transforms the practices of scholarly communication. Thus I focus on the obsession with metrics as a quantification of every aspect of academic labor; so-called questionable academia, that is the massive expansion of questionable publishers, journals, and conferences; following the metrics deployed by institutions, and changes in publication patterns in terms of publication types, the local or global orientation of research, its contents, and the dominant languages of publications.

The book culminates with a concluding chapter in which I investigate whether it is possible to move beyond the inevitability of metrics, and what doing so might imply. I show that the greatest challenge lies in individualized thinking about science and the focus on the accumulation of economically conceived value by institutions in the science and higher education sectors. This is because the problem does not lie in metrics or the power that is manifested in measuring. Rather, the problem is an underlying logic of economization, and it is only by uprooting it that one could change today's academia. Still, any new logic would also be legitimized by new metrics, and so the evaluation game, as a reaction against evaluative power, would continue despite any changes in the ruling logic. Therefore, this book's conclusion is neither a proposal for a "responsible use of metrics" nor a call to abandon the use of all metrics in academia. A third way is needed. Thus the book's key contribution is its call for a rejection of these two potential responses and its insistence on the necessity that we set out now on a course that can offer hope of charting such a third response. It is in this spirit that I sketch out seven principles that I believe should be kept in mind when rebuilding not only a new system of scholarly communication but more importantly an academia that is not driven by metrics.

1

Evaluation as Power

In 1951, Frédéric and his wife Irène were invited to grace the Congress of Polish Science in Warsaw with their presence, as the members of the French delegation. They were among the key officials who had come from abroad. He was the first president of the World Federation of Scientific Workers – the group of scientists who supported communism – and with his wife, had been awarded the Nobel Prize in Chemistry in 1935. Irène Joliot-Curie was the daughter of Maria Skłodowska-Curie and Pierre Curie, who had also been the recipients of the Nobel Prize.

During Stalinism's apogee, the Polish Communist Party decided to reorganize the country's entire scientific landscape and to subordinate all research institutions to the Party. The key means by which they sought to achieve this was by establishing the Polish Academy of Sciences based on the Soviet model. It was during the congress that the framework of this new state institution was presented. The plan was that the Academy replace all learned societies and ministerial institutes as well as most of the research projects conducted at the universities. As the key political institution in science sector, it had to contribute to the Six-Year Plan that was concentrated on increasing the heavy industry sector.

Frédéric was one of the keynote foreign speakers, and he spoke just after the members of the Russian Academy of Sciences. They were all, however, preceded by a speech by a miner and officially recognized "model worker," Alojzy Mozdrzeń, who welcomed the Polish and foreign scientists on behalf of working people. This Polish miner assured the audience that all workers understood that the development of science is based on changing science's program so that it serves the people. Starting with the congress, all research efforts within the Academy had to be planned and coordinated in such a way as to contribute to the economy and – in the case of the humanities and social sciences – to a national culture.

Thus, the Communist Party decided to use its power to reorganize science in Poland, bringing it under its full control. In 1951, the Polish parliament passed the act that established the Polish Academy of Sciences, and one of its titles was devoted to the "planning of research and reporting." From that moment, research projects could be conducted only if they were in line with the Party's plans and implemented by the Academy. Ruled by the Communist Party, the state assumed the control, governance, and evaluation of all research, opting to fund only those fields that contributed to the economy or were in line with the idea of "Soviet man." In this way, a national ex ante evaluation of research was established for all fields. A systematic evaluation of research was carried out, and the key criterion was whether a given institution, group of researchers, or single researcher had contributed to the plan.

Using its power to redefine the meaning of all scientific endeavors, the state thereby introduced new technologies of power through the central planning of research and reporting. These changes transformed science in Poland, as in other countries of the Eastern Bloc, and their effects were evident for many years to come.

* * *

During the Cold War, on each side of the Iron Curtain, a distinct way of thinking about science and its role existed.

The West perceived science mostly as "pure science," that is as an autonomous field organized by an ethos that defended its autonomy against outside influence (Merton, 1973). In 1962, Michael Polanyi characterized science as the republic in which scientists could freely select scientific problems and pursue them in light of their own personal judgments (Polanyi, 1962). Moreover, in this republic, the work of researchers was assessed according to its scientific value that was defined by its accuracy, systematic importance, and the intrinsic interest of its subject matter. In the republic of science, scientists were expected to keep a distance from public affairs.

The (socialist or communist) East, by contrast, perceived science through its social function. John D. Bernal (1939), an Irish scientist who politically endorsed communism, published *The Social Function of Science* in which he presented the idea of science as a tool for supporting a centrally planned society and industry. He considered that the way that science was organized in the Soviet Union was the best model for serving the people, nations, and societal needs. On the Eastern side of the Iron Curtain, the primary function of science was not therefore to cultivate a "pure science" oriented toward solving puzzles but rather to serve socialist society and its economy. Given that it was necessary to reorganize the whole scientific landscape in Poland in order for this to happen, the process was centrally planned and implemented. Science was therefore to serve rather than to build and sustain its own realm.

Today the Iron Curtain is a thing of the past. And yet I agree with Roger Pielke Jr. who, in assessing Bernal's legacy, states that his "ideas on the social function of science have triumphed on nearly every count" (Pielke, 2014, p. 428). While neither the Soviet Union nor the Eastern Bloc exist anymore, at a global level, science is perceived through its social function. For instance, for several years now, one of the key criteria in Western research evaluation systems in the UK, Australia, or the Netherlands is the societal impact of research, which is defined mostly as the effects of research on the economy, culture, society, or quality of life beyond academia (De Jong & Muhonen, 2018; Derrick, 2018). During the Cold War, scientists in the West believed that they had to keep their distance from public affairs. Now, by contrast, they have to contribute to them and to solve problems, which today are called *grand challenges* (Omenn, 2006).

The current system of science funding is hybrid. On the one side, there are competitive grants and block funding for science and higher education institutions. On the other side, various institutions run programs by which specific *societal* challenges like pandemic research related to COVID-19, well-being, food security, and resource efficiency are defined as the goals of research which might be eligible for additional funding. This applies to institutions around the world – from the National Institutes of Health in the US to various agencies in European countries, to China and countries in all other regions. Moreover, global actors like the European Commission also fund research within the lens of grand or societal challenges.

One can observe that while the language with which the role of science is described has changed, Bernal's ideas are still vital. In the descriptions of grand challenges, it is difficult to find phrasing about research needing to serve the people, but one does find that such policy priorities address major concerns shared by citizens. Nonetheless, researchers can still apply for various grants in line with their own research interests, as members of Polanyi's republic of science. At the same time, however, a parallel funding path is becoming increasingly important: that of solving grand and societal challenges through centrally planned research subjects and themes. In either case, the societal impact of research is becoming as important as the scientific value of an investigation's results.

Within these two models, states have played an important role as actors that could either guarantee the autonomy of the republic of science or require that research has a societal (often predefined) impact. For states to achieve their interests, technologies of power are necessary, and one of the key technologies is evaluation.

1.1 Technologies of Power

Evaluation as a social phenomenon drives today's society. It serves to determine the worth, merit, or usefulness of something by reference to a set of values and varied protocols, instruments, and goals that are external both to evaluation itself and to what is evaluated. Dahler-Larsen argues that we actually live in an "evaluation society" (2012) in which evaluation not only describes what has worth or represents value from the evaluation perspective but also describes and constitutes what evaluation claims to measure (Dahler-Larsen, 2015). This observation suggests that in the evaluation process, one does not observe a static relation between those who evaluate and that which is evaluated. Evaluation is rather a dynamic social process of constructing the evaluated objects in and through evaluation itself. In this book, I build on Dahler-Larsen's conception of the constitutive nature of all evaluation processes. This nature implies that evaluation is perceived and used not only as a tool for determining value or merit but also as an instrument of social change across various policy regimes.

Evaluative power is the capacity of the state (or other actors like global organizations) to influence and shape the definition of a key area and to change the behaviors and practices of individuals or institutions by deploying varied technologies. I use this term to name power relations produced by the state and its various policy instruments or, in other words, technologies of power in the science sector. The state's capacity to influence individuals and institutions is mediated by power relations. Evaluative power in science is based on constructing measures and measuring science and research. A key technology that serves to produce and sustain the evaluative power of the state in science is the research evaluation system.

A technology of power is a medium by which the state realizes its interests (e.g., priorities in some research areas) as the owner of public funds. A technology of power is embodied as a set of protocols, metrics, indicators, and policy aims. Like other media, such an embodied medium is not neutral (McLuhan, 1994). This means that through the state's process of constituting this technology, the public sector is influenced, shaped, and potentially transformed. A research evaluation system as a technology of power might also be understood as an "evaluation machine," that is, following Dahler-Larsen's definition (2015), a structure or a function without any subjective or human representation that "lives its own life."

The task that I have set for myself in this book is to show how the evaluation of both the political institutions of the state and the knowledge produced by researchers working in them became an inevitable part of the research process itself. In addition, in this book, I draw attention to the consequences of these processes for academic labor. In defining universities (and at a broader level, all research-oriented institutions), I follow Pusser and Marginson (2013). They understand universities as political institutions of the state because they require state resources, the state provides them with certain benefits, and universities in turn gain some authorities from the state. Both public and private institutions can be understood as political institutions of the state because many governments position universities as part of the state's portfolio of responsibilities. However, given that research, innovation, and knowledge are crucial for the development of states, this relation is complex and goes beyond the mere provision of benefits and authority. From an economic point of view, the science sector is thus a strategic one. According to the Triple Helix thesis that describes relations between university, industry, and government, universities can play an enhanced role in innovation within increasingly knowledge-based societies (Etzkowitz & Leydesdorff, 2000).

Research evaluation systems are science policy instruments that consist of the sets of protocols, measures, indicators, and policy aims that are used for assessing the research productivity and activity of political institutions of the state. Research evaluation systems are some of the key instruments used in various countries, including Australia, Argentina, the Czech Republic, Finland, Norway, Italy, Poland, and the UK. In other instances, they are perceived as performance-based research funding systems or their key constituents.

In the ongoing discussion on research evaluation systems, the systems in these different countries are named as homogenous, examples of the same type of instruments (Hicks, 2012; Zacharewicz et al., 2019). For instance, the Research Excellence Framework (REF) of which the first forerunner was established in 1986 in the UK is indicated as the first research evaluation system, and then other systems are enumerated chronologically. These include, for instance, the Comprehensive Evaluation of Scientific Units in Poland, launched in 1991, and the Research Quality Framework (now replaced by the Excellence in Research for Australia) launched in 2005. Moreover, the advent of the first research evaluation systems is traced back to the 1980s, and their roots are connected to the spread of the tenets of New Public Management, global competitiveness in science, and the knowledge economy (Hicks, 2012). Thus, in the early policy statements presented by governments implementing research evaluation systems, one can find mention of numerous similar themes, including the distribution of state funding, the internationalization of research, and the general pursuit of excellence.

In this book, I want to critically consider the one which is taken for granted in studies on research evaluation and national science policies, that is, the homogeneity of research evaluation systems in terms of the conditions of their formation. While one can identify similar rationales presented in various countries as arguments for establishing research evaluation systems (e.g., funding distribution and the improvement of research productivity), this does not justify putting those different systems into a single category.

It is useful, for the purposes of this investigation, to look into some of the similarities between systems in terms of the protocols, measures and indicators used, and the policy aims. On the one hand, the similarities allow us to compare systems and investigate how they are constructed, including, for instance, what kind of publication counting methods they use. At the same time, such a comparison even allows us to analyze how these systems influence the productivity of universities in terms

of the number of publications. On the other hand, however, at a superficial level, the
similarities conceal actual and deep disparities that are the products of the different
contexts in which research evaluation systems were elaborated. Understanding and
uncovering these differences enable us to show that the current research evaluation
systems vary not only because they use different metrics and set different policy
goals, more importantly, in light of this book's aims, they differ because some of them
were established in the countries of the Eastern Bloc, where the so-called research
evaluation systems had existed before the advent of New Public Management.

At first glance, any two research evaluation systems can appear similar. For
instance, the Australian and Polish systems use similar journal rankings and discipline classifications, and today even societal impact is assessed in a similar way
in these two countries. However, the context and the conditions in which these
systems were established are different.

The Polish research evaluation system was established in 1991 just after the
start of Poland's economic and social transformation from what was later termed
"real socialism" to democratic society and the free market. Therefore, the Polish
system was one of the key policy instruments that served the depoliticization of
research and the implementation of objective measures within the higher education
and science sectors. Given the goals of the process of transformation, establishing a new research evaluation system was imperative as a condition for moving
away from a centrally planned economy. My use of the word "new" for describing
the system is intentional. In Eastern Bloc countries, not only were science and
the economy centrally planned, so too was the research evaluation system which
served to promote the realization of socialist science goals. And yet, despite these
singular characteristics, this chapter in the history of research evaluation and the
measurement of science is all too frequently omitted.

The Australian research evaluation system was launched as part of a five-year
innovation plan called Backing Australia's Ability. This overall state strategy
aimed to build a knowledge-based economy and to enhance the government's
ability to manage the higher education and science sectors in the global context.
The goal of assessing research quality and the impact of research was to provide an answer as to whether public funds were being invested in research that
would deliver actual impacts and provide benefits to society. In this perspective,
it is beyond question that New Public Management is crucial for understanding
the background against which the forerunner to the Excellence in Research for
Australia was designed. Key concerns at the time were public accountability for
resources and the search for the most effective way of determining the allocation
of funding, at the time of budgetary restrictions, to Australian universities. The
context in which the Australian system was designed is similar to that in which
the first and later versions of research evaluation systems were established in the

UK. Tracing this move from accountability for public funds to the assessment of research excellence and the impact of research, one can discern the policy aims that have been prioritized by the state. Still, accountability for public funds and ensuring that they are invested in research that can benefit the wider community continue to be key elements of the environments in which research evaluation systems are designed in both Australia and the UK.

These two examples of research evaluation systems highlight the contrasting climates in which the systems, measures, and metrics were drawn up in Australia (and the UK) and in Poland. More significantly, these diverging contexts have had an important impact on how the same indicators or methods of evaluation come to be perceived differently by the academic community in these countries.

For instance, peer review is usually invoked as the best way of evaluating research results and the impact of research. Peers are treated as the key pillar of science, even though criticisms of peer review are occasionally raised. Thus, the peer review practices implemented within the research evaluation systems in Australia and the UK are presented as benchmarks for other research evaluation systems. In relation to this, the well-worn argument is made: While metrics might be useful, only peers can actually evaluate research and its impact. And yet the key factor in peer review, that is peers themselves, can also be perceived as the greatest weakness of the research evaluation systems, which metrics can be understood as counter-balancing.

In the post-socialist countries of the Eastern Bloc, peers were the hallmark of centrally controlled science: Their decisions were based not on merit but on the political agenda. Wouters (1999) cites an extract from an interview with A. A. Korennoy – a PhD student from Gennady Dobrov and one of the founders of scientometrics in the Soviet Union – who explains that even scientometrics' analyses of research efficiency and productivity were not used to inform policy decisions and funding: "The decisions taken were mostly voluntaristic and guided by completely different considerations. The funds were allocated not according to the front of research but according to personal acquaintanceship" (Wouters, 1999, p. 92). Therefore, one of the ways of making central planning in science a thing of the past during the transformation period of the 1990s was to rely on metrics that were perceived as objective that is not dependent on peers' decisions. Thus, for example in Poland, researchers did not trust their peers and preferred metrics (Mishler & Rose, 1997). Accordingly, each new version of the Polish system was more and more metric oriented. However, neither Polish researchers nor scholars investigating research evaluation systems seem to have noticed that the process of constructing metrics is one in which peers and political agendas are very strongly involved and that ultimately metrics are a hallmark of state power.

Resistance against metrics used in research evaluation systems also differs across countries. In the leading countries for research like Australia, resistance against metric systems takes a different form from that in peripheral countries (Beigel, 2021; Kulczycki, Rozkosz, & Drabek, 2019; Woelert & McKenzie, 2018) because metrics – most often designed by central countries – confirm and legitimize the leading position of those countries. In other words, when metrics are based on data that favor publications in English, it is unsurprising that resistance in Australia or the UK should assume a different nature from resistance in countries like France, Italy, or Ukraine. On this point, I am not claiming that Australian researchers do not resist the use of metrics (Hammarfelt & Haddow, 2018). Rather, I argue that their reaction is shaped not only by science systems and metrics themselves but also by the cultural and historical context.

In this book, I argue that it is not only the case that different sets of metrics, in their implementation, lead to different consequences. It is equally important to consider the process through which metrics are constructed because this process is always situated in a specific context, place, and time. Thus, although two different countries might use the same metrics, they may have been constructed for completely different reasons. Accordingly, in this book, I explore how the state constructs measures and indicators and imposes their use on universities and research institutes.

A research evaluation system is a technology of state power which serves to sustain a power relation built by, on the one side, the state and its agencies and, on the other side, universities and researchers. This technology of power in science transforms the production and communication of scientific knowledge; thus, research practices are influenced by various metrics used in research evaluation systems. For example, using an impact factor for scientific journals can simultaneously encourage researchers to publish in top-tier journals or prioritize quantity over quality in relation to their publications. Investigating the actual (un)intended effects of research evaluation systems is a complex task in which many factors need to be taken into consideration (e.g., gross domestic expenditure on R&D, number of researchers, and policy aims). Existing analyses have revealed many interesting dependencies across funding levels, the metrics used, and the policy aims prioritized. Despite this, many questions related to the rise, development, and role of research evaluation systems remain to be addressed.

1.2 The Evaluative Power of the State

Even authoritarian governments seek international legitimization of their actions. This is why Frédéric Joliot-Curie and Irène Joliot-Curie were invited to participate in, and thereby sanction the event at which the state transformed the academic landscape in Poland. The state had various political and policy tools at its disposal with which to achieve this, all of which were manifestations of its power.

The state has always distributed resources and regulated the public sector. However, the way in which this is organized has changed significantly over the past decades. When, at the beginning of twentieth century, Weber (1978) described how the rationalization of society produced bureaucracy, the characteristics of the public sector were distinct from those of other types of organization. It was only in the bureaucracy that roles were separated from persons, structure organized in hierarchical manner, favoritism eliminated, and a regular execution of assigned tasks implemented. A century later, in many countries, public administration is organized in a manner similar to corporate or private institutions that use key performance indicators, contracts, and a linear model of input–output budgeting, while implementing accountability systems around resource use (Dunleavy & Hood, 1994). This is because, as DiMaggio and Powell (1983) argue, the rationale for bureaucratization and rationalization have changed. Today, the state has become the evaluative state (Dill, 2014; Neave, 2012). The reform of public administration, from bureaucracy to the evaluative state, has been identified with the rise of so-called managerialism or New Public Management (Hood, 1991), which describes a particular way in which relations across government, public institutions, and society have been transformed.

Through varied historical transformations, universities have been confronted with diverse new expectations about their missions, tasks, and organization. The classic conception of the university is most often connected with the idea of the Humboldt University whose structure was defined by a set of autonomous chairs with students affiliated to them. Such universities were autonomous in the sense that professors (chairholders) were autonomous in terms of teaching and research. Since then, the idea of the university has changed many times, and universities have been understood, among other things, as public agencies, corporate enterprises, or innovation-oriented institutions. Autonomy within the university has also been redefined and today one encounters two main approaches: The first one still promotes the autonomy of academic staff members as in the Humboldt university, whereas the second highlights the autonomy of university leaders to define and realize university strategies and to manage the institutions. These approaches emphasize the mission (teaching vs. research) or the autonomy of the university (academics vs. managers). Nonetheless, they do not frame these institutions as implicated within power relations that derive from their dependence on the public funds distributed by the state. And yet today universities and other political institutions of higher education, together with the science sector, are influenced by science policies, including national ones.

As a concept, the evaluative power of the state can be used in many forms and contexts. Using it productively requires that one both identify and prioritize key aspects of power. Even then, however, given that the concept of power is a very complex one, the likelihood remains that one is charged with being imprecise or unclear. In this book, my focus is on investigating how the power of the state transforms the production of scientific knowledge and the very research practices themselves. As I argue in Chapters 2 and 3, this power impacts on political institutions and researchers both directly and indirectly. In terms of its direct impact, this is exerted through state regulations, policy documents, and policy decisions. Its indirect impacts are realized through the shaping of the conditions and environments in which researchers work, which include labor conditions, types of employment and contracts, methods of assessment of academic staff, and the amount of financing distributed to universities.

The key causes of these indirect impact are the technologies of direct impact, that is the above-mentioned regulations, documents, and policy decisions. Nonetheless, both kinds of impact produce both intended and unintended effects simultaneously. In other words, the technologies of direct impact can influence the productivity of some researchers, bringing them into line with policy aims (e.g., the increase in publications in top-tier journals), while at the same time, another group of researchers can transform their publishing practices in unintended – from the science policy point of view – ways (e.g., more publications in local scholarly publication channels). Additionally, the shaping of work environments as the effect of indirect impacts can improve researchers' productivity and their focus on the societal impact of research which might be an intended effect of policy regulations. However, the indirect impact of, for instance, exactly reproducing national evaluation procedures at the university or faculty level can lead to a deterioration in the quality of academia as a workplace and, as a consequence, reduce the innovativeness of research. Such an effect would be unintended from the science policy point of view.

In order to investigate the effects of state power, we must go beyond the assumption that state power is a very complex mechanism embodied in power relations between the state, political institutions, and the researchers working within them. We must also view it as a set of actions and state capacity as mediated and implemented by numerous technologies of power or policy regimes. Hence, in examining the power of the evaluative state, it is necessary to combine two – at first glance – antithetical perspectives: Foucault (1995) and Lukes (1974)' definitions of power. These two conceptions can be perceived as antithetical because while for Foucault, power is an unintentional and overarching condition, for Lukes, power is a person's capacity to influence or change the interests of someone else. Lukes' definition of power as a capacity to do something highlights its intentional dimension. Whichever definition one adheres to, the following question needs to be addressed: Can the power of the state be both unintentional and at the same time, intentional? If one defines power as power relations and technologies and then focuses on the effects of the use of these technologies, one finds that in order to understand power itself, one must understand the (un)intended effects produced by power technologies. One should therefore conceive of power as both an overarching mechanism and as the capacity to change someone's interest and action; doing so enables us to investigate power effects in a holistic way.

Foucault (1995) argues that power designates the complex and all-encompassing condition that produces and shapes our social reality, actors, objects, and relations. Understood in this way, power is not intentional action or strategy: It is because of its complex nature that power can change us, and not because some actor or institution that "has power" decided to do so. Thus power is not something that actors can have but is instead a complex relation in which they are involved. People and institutions involved on a continual basis in such situations internalize external control that makes them more or less willing to subject themselves to the societal norms and expectations that are the product of power relations.

Through its varied technologies, power colonizes people's minds. It is disciplinary power that becomes embedded in the various administrative routines of institutions. As presented in *Discipline & Punish*, disciplinary power is connected with closed spaces like prisons, hospitals, or schools in which control was exerted together with the restriction of freedom. In the era of New Public Management, however, these institutions have been transformed. Therefore, the logic of power itself has also been changing: It no longer functions under this disciplinary modus operandi but rather relies on the semblance of freedom coupled with uninterrupted control. In the books, he wrote after *Discipline & Punish*, Foucault argued that it is not possible to study the technologies of power without also considering the political rationality that underlies them. Thus he coined the concept of "governmentality," which combines the perspective of the state that governs others and the perspective of the selfthat governs itself (cf. Lemke, 2002). In this optic, subjects treat external norms as their own and govern themselves in order to meet external expectations. This observation is critical in my consideration of scholars' attitudes to various science policy instruments and their resistance to power structures, which are presented in the next part of this book. Although power's effects cannot be resisted, subjects are nonetheless aware of power relations and technologies and can, at least hypothetically, try to resist power.

Power, according to Foucault, is an unwilled complex mechanism that affects individuals and institutions. Steven Lukes, with his radical view of power (1974), defines it in a different way. Lukes shows that in the past, power was mostly understood as a one- or two-dimensional capacity, and his argument is that it should instead be defined as three-dimensional capacity. The one-dimensional view of power defines relations between persons or institutions as the capacity to convince a person to do something which they would normally not do. The two-dimensional view redefines relations and emphasizes the idea that having power means having the capacity to put up obstacles and, in this way, reduce and control others' options. As a way of developing these two approaches, Lukes suggests that one characterize three-dimensional power as a person's capacity to influence, shape, or change another person's core interests.

In this view, then, power manifests itself through domination, that is acts of influence and manipulation. A person who influences or further, alters the interests of another person is an influencer who is working to promote an agenda. To put it differently: Power through domination is an intentional action taken by a specific person, by people, or institutions. While an influencer might not be recognized as such by those who are being influenced, it is still possible to reveal the power relations between them. In this way, Lukes' (1974) approach, contrary to Foucault's, highlights the intentional dimensions of power and focuses on decision-making and control over the political agenda (p. 25).

As a complex apparatus, power requires structures that need to be sustained. In societies, power's close relationship to knowledge is key to its reproduction. Foucault defined this relation through the concept of Power/Knowledge, where knowledge and information refer to individuals, groups, and institutions. It is the collection, archiving, and analysis of such knowledge that allows power to sustain and reproduce itself. As Weber (1978) showed, collecting and archiving information is one of the key characteristics of bureaucracy. In this way, bureaucracy and – for the past decades – evaluative states use collected and archived knowledge to control and govern, among others, the public sector. Thus knowledge and information become power technologies of the state.

Building on Lukes' conception of power, I define the evaluative power of the state as the capacity to influence or transform the interests of individuals or institutions and to modify their practices and behaviors. Evaluative power is reproduced by the very context that it itself produces. Thus, its capacity is realized not only through the implementation of policy instruments but also through the redefinition of the context. The concept of the evaluative state as described above underlines the fact that the state produces power relations by implementing various evaluation instruments. However, in this conception, the emphasis is mostly on the intentional actions that influence (state administration, policy makers). It is my contention that by bringing together Foucault and Lukes' approaches, we can deepen our investigations of the (un)intended effects of research evaluation systems.

From Foucault's perspective, disciplinary power enables the sustenance and reproduction of all-encompassing power relations; however, following the logic of this model, it is not possible to identify actual agents or influencers, as disciplinary power influences and transforms individuals and institutions themselves. One can therefore conceive of the inevitable and inescapable evaluation of science by and, as a consequence, governance of science by the evaluative state as a form of discipline in Foucault's sense. Evaluation, like discipline, is based on normalization and constant surveillance and drives both individuals and institutions to continuous self-evaluation, and to comparisons between themselves and other self-evaluating entities that are also subjected to this all-encompassing mechanism.

In my investigation of the evaluative power of the state, I build on the idea that the intentional dimension of the power (in Lukes' sense) of the evaluative state needs to be examined in combination with power's unintentional dimension (in Foucault's sense), in which evaluation is understood as a form of discipline. In this book, my core concern is with the concept of evaluative power rather than with that of the evaluative state. This is because I am interested in the following two areas: (1) how the state transforms scholarly communication by constituting evaluative objects and by evaluating political institutions and (2) how scholarly communication is transformed by various self-evaluation practices (resulting from the internalization of evaluation norms) and forms of reactions and resistance against evaluation itself. While the concept of the evaluative state highlights the capacity of the agent, that is, the state, to manage and govern through diverse evaluation regimes, the concept of evaluative power focuses on power relations (and not the agent's capacity) that are cocreated and mediated by the agent and its technologies of power.

If a person is in the position to decide whether to measure, this implies that they hold power. This power might also be strengthened if they can also determine the way in which that measuring occurs. In publication-oriented academia, the measure is well known: it is the publication itself as characterized by various numbers such as the number of citations or social mentions or the opinions of peers and experts. However, as I explain below, measuring is not the measure, and measuring is linked with deciding how this measure is used depending on "what" and "who" is measured.

Witold Kula (1986), in his *Measures and Men*, reconstructed the social processes involved in constituting varied measures and ways of measuring. In feudal society, there was the widespread view that it was legitimate for a tradesman to use one measure when buying and another when selling. However, even one measure could have two different types of use. For instance, when a merchant was selling a bushel of grain, the bushel was struck (strickled) or the grain would be leveled with the bushel's rim. Yet when someone repaid the grain to the same merchant, the bushel needed to be heaped or "topped up," simply because the merchant had power to enforce this (Kula, 1986, p. 103). Here then is an example of the use of a single measure (the bushel) in which the quantity of grain differs because the power relations are different. Thus one can say that power manifests itself in the imposition of a method of measurement.

Evaluative power in science is embodied, among other things, in research evaluation systems, funding agencies, and varied accreditation procedures. Where the state has public control over institutions, there always exists some form of evaluation which is inevitable and inescapable because of the very nature of the evaluative state (Neave, 1998). In the science sector, evaluative power designates the power relations produced by the state across political institutions, researchers, and state officials and policy makers. These power relations manifest mainly in (1) the design of measures, (2) the use of these measures to evaluate political institutions and researchers and to make a range of decisions using the evaluation results, and (3) the reactions and resistance of researchers against evaluative power and its effects.

These three manifestations of power relations determine the three main lines of inquiry pursued in this book: national science policies, research evaluation systems, and the evaluation game. The intertwining of these areas produces tensions across all parties implicated in power relations: the state, academia, and researchers. These tensions are the basis for resistance against the imposition and use of measures to evaluate scientific work. Finally, these tensions produce the evaluation game in which the rules and stakes revolve around measures and measurement

1.3 Games as Redefined Practices

In order to investigate the practices of any group or community, we must determine how certain actions can be identified as actions of the same type. Doing so allows us to pinpoint why specific activities carried out by different people should be perceived as actions sharing a common denominator. In other words, to specify why these actions constitute a social practice and how the meaning of this practice and action is reproduced in society.

Science is a cultural practice and as such it consists of rules, norms, values, standards and, in a more general sense, knowledge that are shared by members of a given community. Thus actions can be understood as realizations of a given practice (e.g., writing this book as a practice of scholarly communication) when a person follows specific rules (values, standards etc.) shared by a community of scholars (e.g., a manuscript should present the research results, and relevant works from the field should be cited). Action alone or even many actions alone do not constitute a practice because every practice is oriented toward interpretation. For example, this means that the specific action of writing this book is a realization of a practice of scholarly communication (i.e., its meaning is determined by a given practice) only when other researchers can interpret my action (writing a book) and its results (this book) in light of the rules and values shared by researchers. Practice is meaningful when either its actions or its results are communicated and accessible to other members of the community. For instance, if novel and clear argumentation in scholarly work are important values for researchers, then researchers who read this book can interpret my writing and assess whether the rules, norms, and standards have been properly followed in the attempt to realize these values. However, an unpublished book is an output of writing action but not of the practice of scholarly communication because members of the academic community cannot assess and interpret it.

Researchers are always involved in various scholarly practices at the same time. They do research, write papers, analyze data, evaluate proposals, manage institutions, organize conferences, communicate with peers, and realize countless other practices to which they are socialized by taking actions and by experiencing their effects through feedback. The meaning of their practices and actions is grounded in values that determine what the best path is for realizing a given value. In other words, what norms one should follow and according to what rules one should act. In practicing science, however, researchers often have to assume a dual identity or dual loyalty because of conflicts between the values that drive their actions.

Researchers have to realize numerous practices that are specific to the institutions in which they work or to the scientific discipline to which they belong. Loyalty to the institution in which they work is always a local form of loyalty, but loyalty to their discipline is always global because, by its very nature, science is international. Therefore, a point of reference for assessing the value and meaning of someone's actions (e.g., writing and publishing a paper) can be set either locally or globally. This implies that some actions can be in line with the values shared by researchers employed in a given institution and simultaneously, out of step with the values shared by researchers in a given discipline. For instance, because research is international, within many disciplines the best (or even the only) way to communicate research results is by publishing them as a journal article in a top-tier journal. This is a practice grounded in the value of promoting the broadest possible communication with peers around the globe. This value of global communication is set for all researchers and all actions taken by them, that is their publishing activity, are interpreted in light of this value. Nonetheless, from the second half of the twentieth century in various European countries and in the United States, one can encounter the practice of publishing a *Festschrift*. Let us take a look at this practice to see how conflict between loyalties can occur.

A *Festschrift* is a scholarly book (most often an edited volume) honoring a respected scholar and published during his or her lifetime. Editing a *Festschrift* or contributing to it (by writing a book chapter) is a way in which colleagues, former students and friends can pay homage a researcher. Most of the time, a *Festschrift* consists of original contributions prepared especially for the book, although occasionally, contributors may submit papers that are difficult to publish elsewhere. The practice of contributing to a *Festschrift*, which is often published by a publisher with local distribution only, produces a tension between the global and local loyalties of researchers. On the one hand, to write a book chapter is to go against the standard practice of the best way of communicating research in one's discipline. On the other hand, writing a book chapter for a *Festschrift* is an appropriate way of cultivating values shared by colleagues from an institution. In the course of everyday academic work, all researchers face such dilemmas and tensions, because their working conditions are shaped mostly by the local context in which their institution operates. And yet recognition of their work by their disciplinary community is grounded not in their local, but rather in global terms.

Given that their actions are driven not only by a desire for recognition but also by the need for stable and healthy working conditions, researchers act and practice under parallel (and sometimes mutually exclusive) value systems that produce multiple tensions between their local and global identities. Such tension between two loyalties can also be understood in the light of the concept of the evaluation gap described by Wouters (2017), that is, as tension between what researchers value in academic work and how they are assessed in formal evaluation exercises. There is no indicator, as Dahler-Larsen (2022) argues, that could finally close this gap. However, the tensions between the values of an academic community and those grounding evaluation systems are not the only ones that affect researchers working in academia.

Academia is further subject to tensions generated by power relations across the global and national planes, and between institutions, decision makers, and researchers. When a state produces national regulations for research evaluation, all of academia in that country needs to situate itself within this new environment that has been produced by evaluative power. In sum, while the state evaluates, academia is evaluated and, on being evaluated, researchers react.

It is not, however the case that academia only follows state regulations (e.g., collects and archives information, calculates statistics, and assesses researchers). It also reacts to evaluative (disciplinary) power through various forms of adaptation, resistance, and struggle. As Foucault argues: "people criticize instances of power which are the closest to them, those which exercise their action on individuals" (Foucault, 1982, p. 780). In the case of research evaluation systems, this manifests itself in the fact that people in academia focus their criticism on the state administration and burdensome nature of reporting about their work, rather than the basic conditions that allow evaluative power to arise. Among such conditions, I include the rationalization of society that fed into New Public Management, academic capitalism, audit culture, and the neoliberal university. These necessary conditions for the existence of the evaluative state produce an all-encompassing mechanisms of power. Thus, while they are invisible for those in academia during their regular work, discussions, and reflections on academia itself, their consequences are nonetheless felt at work.

It is only through critical investigation aimed at uncovering power relations that one can elucidate the fact that any effort to change academia needs to take into consideration not only current technologies of power but also the conditions that make those technologies possible. While revealing power relations is one of social research's critical tasks, it should not be its ultimate objective. Investigators focused on research evaluation should go further and recommend next steps that can deepen our understanding and improve the situation of all (or at least, some of) the parties implicated in power relations. Doing this, however, requires a better understating of the reactions, responses, and resistance that surface within academia when it operates under evaluative power.

In this book, I argue that resistance in academia against evaluative power manifests itself through various forms of the evaluation game. Evaluative power is reactive because it causes those working in academia to think, act, and react differently (cf. Espeland & Stevens, 2008). I use the term "game" because evaluative (disciplinary) power produces an all-encompassing situation that is nevertheless rule-based and has defined the ends. Moreover, I build from Foucault's idea that relationships of power are "strategic games" between liberties in which some people try to determine the conduct of others (cf. Lemke, 2002).

In defining the game, I construct a framework in which players (e.g., political institutions, researchers) are socialized for the game by taking actions (e.g., writing manuscripts, planning research) and by experiencing their effects through feedback which is deliberately built into and around the game (cf. Mayer, 2009). In terms of the rules of the game, these are set by those who have power in the power relation. The idea of game is one of ways of conceptualizing social interactions in any environment in which rules are explicitly stated and in which rule-makers might, to a certain extent, be identified. This approach has a long tradition in social sciences and one can point to many different perspectives, among them, George H. Mead's (1934) interactionist approach, Thomas S. Szasz's (1974) perspective on games as a model of behavior, and Erving Goffman's (1972) approach in which the game is the context in which behavior takes place.

Within studies of academia, "playing the game" is not a pejorative term akin to the idea of "gaming" but rather a name for the day-to-day practices of academic labor within a rule-based environment. Bourdieu uses the concept of game to explain the meaning of the field and argues that a game has no explicit or codified rules (regularities) but is worth playing (Bourdieu & Wacquant, 1992). In writing about evaluating the evaluation game, Elzinga uses the term *evaluation game* to analyze a methodology of project evaluation and writing of the possible effects of evaluation on research practices. Kalfa et al. (2018) using Bourdieu's understanding of game explore how academics respond to managerialist imperatives within the academic game. Lucas (2006) describes research in the competitive global market as the *international research game*. Fochler and De Rijcke (2017) use the term *indicator game* to name the "ways to engage with the dynamics of evaluation, measurement and competition in contemporary academia" (p. 22). Blasi et al. (2018) argue that evaluation can be treated like a game played by the institution against the actor: "the actors receive a payoff from their behavior according to predefined rules and will engage in the strategic games to beat the rules. The extent to which strategic games can be played depends on the nature of the rules" (p. 377).

Yudkevich et al. (2016) use the concept of the game as a framework for their edited book *The Global Academic Rankings Game* and define the *rankings game* as a high-stakes exercise that exerts influence on institutions' internal policies.

Suits (1967) demonstrates how games are goal-directed activities in which inefficient means are rationally chosen. For instance, while playing soccer, no one can touch the ball with their hands unless they are the goal-keeper. When a player is kicking the ball toward the goal and another player (not the goal-keeper) decides to use a leg instead of a hand to block the shot (although the latter would be much more effective), then he or she is rationally using an inefficient means which is permitted by the rules. Thus, in all games, we need to know what means we can use and what means would be classified as rule-breaking.

How do games – which are also social practices – differ from other rule-based practices? Suits (1967) argues that people obey the rules simply because such obedience is a necessary condition to make *playing the game* possible. In other words, following the rules makes it possible for the game to take place and thus we follow the rules for this purpose. In other types of practices and activities, there is always another (external to the game itself) reason for conforming the rules. If we consider, for example, moral actions, following the rules (e.g., rules derived from religious values) makes our action right and not following the rules makes our action wrong. An analogues situation can be found in communication acts. If, for example, the aim of the act is to respect, through mourning, those who recently died, people can follow the rules and engage in a moment of silence, which then makes the silence a gesture of respect. Not following this rule when others are makes our action (e.g., speaking loudly) wrong in the context of this specific communication practice.

A social practice – such as language – cannot be created in a short period of time by decision or by the act of one or a few persons. For instance, researchers write and publish journal articles because in their disciplines or institutions this is how science has been done for years. It requires a great deal of work and time to change such a practice through a bottom-up approach, which is to say by common decision of the practitioners themselves (e.g., researchers from a given institution who practice research). This would apply, for instance, to a desire to shift the practice in order to start publishing more internationally oriented papers or to publish in English when an institution is based in a non-English–speaking country.

One cannot reduce the process of altering a practice to the presentation of new aims and goals (e.g., "our institution needs more publications in English") and implementing new techniques (e.g., "researcher can attend English lessons"). Such a process requires a change of mentality or rather of the collective representations shared by a community and embedded in their practices. When an institution or discipline decides to modify a practice, it initiates a process that consumes a great deal of time and work.

A change (even a radical one) of practice or the implementation of a new practice can also occur fairly rapidly through top-down approaches, that is through the implementation of a new set of rules or by forcing a change in the current situation. For instance, this can occur when a state which, for several years has been evaluating institutions according to the number of peer-reviewed publications they produce, informs institutions that from now on, they will be evaluated and financed according exclusively to the number of peer-reviewed publications they produce in English. In such a case, one can say that the norms, rules, and values of academia have been subjected to rapid transformation. From the perspective of researchers and institutions, rule makers introduce new regulations as if they were pulling instructions from a new boardgame box: from now on you have to follow these rules because you work in a political institution of the state. In such cases, the implementation of new rules by evaluative power will appear abrupt, even if there has been a process of public consultation, or of preparing the new regulations. The outcome of such processes is that a completely new situation is produced.

Finding themselves in a new situation, researchers and managers have to gage how to act in order to comply with the criteria and values defined by the new rules. They do not only look for the most efficient way of practicing science or communicating their research but also start to think about what means they need to use so as to secure their position in the new (evaluative) situation and to follow the new norms. If in a new regime of rules, only English language peer-reviewed publications count, then researchers, for example, those whose main disciplinary language of publication is German, will start to evaluate how to go on with their work and publishing. In this instance, according to the values shared within their discipline (discipline loyalty), the best way to publish is to publish in German. And yet according to the values shared by managers from their institution, it is best to publish in English or in both English and German. In this way, that is by a rapid implementation of new value regimes, multiple tensions between researchers' loyalties are created. If I am a researcher who needs to decide how to act in such a situation, I can choose one of the following strategies: (1) I am loyal only to a discipline and publish only in German; (2) I am loyal only to an institution, and I stop publishing in German and start publishing in English; (3) I try to be loyal both to a discipline and to an institution, and I publish both in English and German, even though publishing in German is perceived as a waste of time and resources.

Researchers and managers are regularly confronted with such decisions. In this way, academic work becomes a game, that is, a situation that is produced by a top-down implementation of new rules related to a social practice that has been cultivated for many years and has established norms, values, and rules. Introducing new or modified rules initiates the process of institutionalizing the practice and establishing the game. When a game coheres, researchers and managers start to play; each of them has their own individual strategy, yet at the same time, interactions between players can also modify the way in which they play.

A change of rules provokes researchers and managers to start searching for ways of engaging in practice (in Suits' sense), which would allow them to follow the new rules at the lowest possible cost, because the rules can be changed again at any moment. This redefined practice is realized by bringing actions into line with the new rules, which does not necessarily mean that they are in line with the values and aims of the institutions, disciplines, or even the state and rule-makers. What matters is that one meets the criteria laid out by the new rules, whatever the cost. And the only reason for conforming to the rules is in order to reproduce the situation in which one can follow the rules, because reproducing them is what allows one to be employed in a given institution. This is why I call the *game* a specific form of social practice: It results from a rapid top-down redefinition of ongoing practices by the implementation of new (or modified) rules through various technologies of power (e.g., research evaluation systems).

The game is not a typical bottom-up social practice; it is rather a top-down redefined social practice. The power to rapidly redefine a social practice is always external to practitioners. Practitioners become players because they want to at least maintain their current position. They play because there are resources at stake in the game (e.g., stable work conditions, funds for research), with the goal of game being to win and gain resources. As long as an evaluation system is operational, the evaluation game does not end. While the stakes of the game vary, depending on various conditions, with players having different starting points, the key issue is that once the game is established, those working in academia are forced to play it. Therefore, they try to adapt and modify their practices in line with the new rules and aims implemented by the evaluative state.

An adaptation constitutes a *strategy* as to how to play the game and, in this way, scholarly practices become evaluation-driven practices in academia. A strategy is not something that is intrinsic to a game but rather something that a player brings to the game. As Avedon (1981) argues: "it is something that the player develops, based on his past experience, knowledge of the game, and the personality of the other players" (p. 420). Thus, different strategies are used in different institutions, fields, or countries. People in academia can either adapt (or adjust) to the system, try to ignore it, or try to change it (cf. Bal, 2017).

1.4 Defining the Evaluation Game

Weigl's situation, described at the beginning of this book, can serve as an example of the evaluation game: He published a number of papers in order to hold on to his scholarship. This action was not in keeping with his loyalty to his discipline, but he was compelled to take it by the all-encompassing situation that defined the conditions of his work.

The evaluation game is a practice of doing science and managing academia in a transformed context that is shaped by reactions to and resistance against evaluative power. Such a game is established in a dialectical process: Through evaluative power, the state introduces new rules and metrics, while researchers and managers in academia devise various strategies for following these rules at the lowest possible cost. These strategies – as forms of adaptation, response, and resistance against evaluative power – are reactions to new rules and metrics.

Playing the game does not interrupt social practices (e.g., communicating research results) but instead puts the accomplishment of the new goals and compliance with the new rules in first position. In this way, people may start looking for ways to get around these new rules according to which a practice is evaluated. If, for instance, an institution assesses an individual scholar based on the number of journal articles they published in a four-year period, then playing the game would involve an artificial increase in the total number of co-authored articles. Thus, two scholars who used to publish single-author articles could decide to start writing joint articles (or even to add each other as authors to papers actually written by only one of them). This would then occur not because they had started to collaborate, but in order for them to increase the number of articles to their names. Here the game – that is, reaction to the rules of evaluation – has just started. A practice of scholarly communication is still being cultivated (journal articles are published), but this practice takes the form of a game in which the end is not to communicate research results in the best way (from a disciplinary perspective) but rather to communicate the research results in line with the evaluation criteria and with the scholar and institution's interests.

The evaluation game manifests in the day-to-day work of all those involved in the diverse power relations of evaluation. Therefore, the game can be played by all parties in those relations, that is, (1) researchers, (2) managers, and (3) the policy makers who design the measures used within research evaluation systems. Let us consider an example to see how individual actors participate in the game.

Global competitiveness for key resources like funds, researchers, and students, as well as university rankings, exerts a pressure on governments and policy makers to improve the productivity and quality of research in terms of bibliometric indicators. As a rule, bibliometric indicators are only a proxy for research quality, and yet they are identified as the information that shows how well a given country, institution, or researcher is performing. Thus, such indicators play an important role in rankings, and in this way, global competitiveness contributes to the emergence of the evaluation game.

In this context, policy makers start to play. The aim of the game is to boost the position of universities in their country in rankings that are considered important within that country. From the perspective of policy makers, the goal of the game is to legitimize the funding of the higher education and science sectors. It might be the case that policy makers decide on certain areas within the ranking criteria which they believe can be improved, and then focus only on these. For instance, in some rankings, only publications from the TOP 10% of top-tier journals in international databases like the WoS or Scopus are counted. Therefore, policy makers may respond by creating a research evaluation system for assessing and funding institutions. In such a system, publications in TOP 10% would have substantial weight in relation to other publications from outside of them, or indeed, other types of output. The policy makers' response is shaped by the global context of competitiveness and the expectations of the key stakeholder, that is, the government, that institutions from a given country should improve their position in the rankings. By designing an evaluation system, policy makers meet the expectations of ranking criteria. However, the effort focuses only on bibliometric indicators that do not capture the larger complexity of doing research. And yet in this scenario, policy makers have responded to a challenge at the lowest possible cost, that is they have focused only on what is counted and what might be improved in order to increase the position of institutions within a specific ranking.

In such a situation, how might managers of academic institutions react? They might play the evaluation game created by ranking pressures by applying the rules of institutional evaluation at the local level of researcher assessment. This practice is well documented in research evaluation studies and is called "local uses" (Aagaard, 2015). Managers may just copy the regulations, which is the move that entails the lowest possible cost. Thus they may permit researchers, as part of the individual researcher evaluation exercise, to report only those publications from TOP 10% journals.

In this example, the context in which researchers work has been redefined in a top-down manner. Researchers would know that in order to receive a positive evaluation in the upcoming assessment of their work, they would be expected to publish only in TOP 10% journals. In response, the strategies for engaging in the game would vary depending on the discipline in question. For instance, researchers from fields in which scholarly book publications play an important role might stop publishing books and try to redefine their research topics in a way which allowed them to publish in journals. Through such a strategy, publication patterns and publication channels might be transformed. In the disciplines in which publishing in journals is a common practice, researchers might decide to put more effort into publishing in better – from an evaluation game perspective – channels, or they might start to consider how to game the situation by publishing more co-authored publications that are not based on real cooperation.

There is naturally a great diversity of ways of playing the evaluation game. One might consider the consequences of choosing only TOP 10% journals in relation to publishing in national languages, given that the majority of top-tier journals indexed in international databases publish only in English. I will present a detailed analysis of the different types and forms of games in Chapter 6. Here, however, let us summarize and pinpoint what this example tells us about the evaluation game as played by all parties implicated in this particular set of power relations.

As can be discerned from the above example, researchers play the game because a top-down redefinition of the context in which they work has occurred. When one analyzes how the evaluation game is played, one needs to take into account the fact that this game is constituted at all levels (global, national, and local) and that its actors represent all parties within power relations. Consequently, because the game is a social practice involving actors who follow (or not) the rules, and play for certain stakes, the game cannot be reduced or abstracted from the actors who play it, nor presented as existing at a separate level. Rather, one must view all these elements and parts of power relations as mutually constitutive. This is because the evaluation game is a response by all actors within the science system to power relations that are generated by the evaluative power of the state. This state power is, moreover, significantly influenced by the global context of doing and managing science; thus the higher education sector is shaped by global actors and institutions. In other words, one might say that when context is redefined through a top-down process, everybody plays the game. However, it should be noted that researchers, managers, and policy makers play the evaluation game while at the same time playing other games in academia that are related to teaching, their careers, and relations with peers.

1.5 Factors Contributing to the Evaluation Game

One can identify three main factors contributing to the emergence of the evaluation game in science. The first is related to the use of measures (quantitative indicators or metrics) to control, govern, or modify social behaviors in line with particular aims or targets that are external to those being controlled and governed. In such situations, an indicator becomes the target and in this way, the stakes of the game are changed. The second relates to changes in the context in which researchers and managers work due to the implementation of a research evaluation system or modifications to it. The third concerns tensions between the rationalities (or logics) adopted by designers of research evaluation systems and the rationalities actually used by people in academia. Let us look into these three factors in detail.

The first factor contributing to the emergence of the evaluation game is linked to the very nature of any measure deployed to monitor, control, govern, or modify social behaviors. Among other things, measures are used to provide an intersubjective assessment, and yet when they are used, they cause those working in academia to think and act differently (cf. Espeland & Stevens, 2008). Donald T. Campbell made the observation – widely known as Campbell's Law – that any indicator used for social decision-making will become a poor indicator because of its very nature: "The more any quantitative social indicator is used for social decision-making, the more subject it will be to corruption pressures and the more apt it will be to distort and corrupt the social processes it is intended to monitor" (Campbell, 1979, p. 49). Similar research into the repercussions of using indicators, like Goodhart's Law, the Lucas critique or the Cobra effect, highlight what happens when numbers, quantification, and measuring are used to control behaviors and social actors: The indicator itself will become the target, and people will do what is being measured and stop doing that which is not being measured. In his *The Tyranny of Metrics*, Muller (2018) provides numerous examples of the consequences of using indicators in various sections of the public sector such as health, education, and higher education. All of these examples make it evident that if one decides to use numbers (indicators) to assess or monitor social practices, the practice itself will change. However, whether the change will be positive or negative for those subjected to these indicators is not something that can be determined in a top-down manner.

The second factor contributing to evaluation games in science relates to the influence of evaluative power on changes to power's all-encompassing mechanism and the day-to-day practices of academic labor. A change may consist, for example, in the introduction of a new way of reviewing publications submitted to a national evaluation exercise or in an increase in the number of publications that researchers need to present within specific time periods. In either situation, researchers and managers need to decide how to react. In an ideal world, managers would communicate changes in the regulations to researchers. Then, they would have a discussion on how to tackle the change, how it might influence the researcher's work, and what adaptations would need to be made in order to prepare for the change. In actuality, however, changes in evaluation regulations are often rolled out through long and drawn-out policy processes.

It is also the case that even the best academic managers are sometimes helpless in the face of vague policy documents and regulations: It is often difficult to understand how a minor change in regulations might actually influence the day-to-day practices of researchers or managers. Will it increase the burden of administrative work? Will the different scope of information to be collected in order to comply with the evaluation criteria require a new workflow? How might the implementation of a new bibliometric indicator (e.g., the Hirsch index) within national regulations impact on internal evaluation procedures at our university? Such questions follow each modification of the all-encompassing nature of power produced by evaluative power. When a change is made, then those working in academia have to operate in an unknown situation. Confronted with this situation, they start to think about how to continue their day-to-day practices in the same manner as before and to change them only when it is unavoidable or absolutely necessary. Day-to-day practices in academia (defined in terms of "games") are usually long established, but a change in the rules can modify the stakes of the game (e.g., employment stability or funding for research). Moreover, a change in the rules can put the spotlight on the fact that that while those in academia all have to participate in the same game, they do so from different starting points, with some starting points, for instance, those of early career researchers, constituting a disadvantage.

The third factor leading to evaluation games in science consists in the tensions that exist between the different logics that motivate system designers and those working in academia. Those who design research evaluation systems have to make multiple decisions related to the following questions: What will be measured? (e.g., what kinds of activities will and will not be measured); who will do the measuring? (e.g., peers, stakeholders, and experts external to academia); who will be measured? (e.g., will outputs produced by PhD students be evaluated); how will things be measured? (e.g., should qualitative, quantitative, or both methods be used); what criteria will be used? (e.g., do only peer-reviewed publications matter, or are publications for the general public also important); when and how often should evaluations take place? (e.g., annually or once every four years); how will the information collected be used? (e.g., only for evaluation exercises or also for other administrative purposes); how will the results of the evaluation be presented? (e.g., through the ranking of institutions or only through information on positive/negative results); and finally, how will the results be used? (e.g., only for block grant distribution or perhaps also in order to change human resources policy).

While taking decisions on these questions, system designers necessarily adopt certain epistemological assumptions about the cognitive responses of those in academia who will be subject to the research evaluation system (cf. Pollitt, 2013). In other words, system designers assume that researchers and managers behave according to certain logics and in this way, they predict how researchers and managers will behave and modify their practices in response to the roll out of a research evaluation system. In his description of the logic of performance management regimes, Pollitt (2013) shows that two types of logic can be adopted by system designers: (1) a goals-oriented logic and (2) a logic of appropriateness.

According to the first logic, academics and managers are primarily motivated by wages and labor security. Thus, in the course of their daily work, they conduct research while at the same time trying to achieve the goals presented to them by stakeholders or managers. In this way, through the setting of targets, policy makers govern academia (cf. Bevan & Hood, 2007). Within the second logic, academics and managers follow the collective values and collective representations shared by their community. If policy makers want to govern academia, they need to not only set targets but also remodel the environment in which academics and managers work. Such changes can then modify the set of values and collective representations shared by members of a given academic community. It is important to note that these two logics have substantially different consequences for the design of research evaluation systems. If designers assume the first logic as primary, then the system has to have clear goals, targets, and indicators that should be directly communicated to the evaluated community. In keeping with the second logic, academia also can be governed, but this requires more complex agenda-setting, time, and resources.

While these logics are adopted by policy makers and system designers at the macro level, system users, that is, researchers and managers, behave and practice at the micro level. Here, a much greater variety of logics are deployed, depending on the specific time and context (e.g., whether researchers work at a top research-intensive university or at a small local university). Pollitt terms such micro-level logics "alternative logics" (2013). Alternative logics can be adopted by single researchers, groups of researchers, or by the whole community at a university. Moreover, researchers or managers can follow diverse alternative logics at the same time. Pollitt argues that these alternative logics are used by actors who are subjected to performance management regimes. However, some alternative logics are in fact also adopted by policy makers and the designers of research evaluation systems. This is because they have to make assumptions about how academics and managers behave in their day-to-day work and are themselves also engaged within the wider system. That system involves the policies that are realized by the state, as part of which varied logics about people's behaviors are also adopted (e.g., how policy makers and policy system designers think or should think about social interventions).

Pollitt uses the term "alternative logic" to name the mostly unintended effects of performance management. Some of the examples of alternative logics he offers are the *threshold effect* (a minimum target can motivate those falling below the target but also de-motivate those who have already performed above the target), the *ratchet effect* (managers may be tempted to hit but not exceed the target if next year's targets are based on last year's performance), and *cheating* (not bending but breaking the rules). In other studies of performance management, one also finds diverse alternative logics that are also termed the "unintended consequences" of performance management (Smith, 1995) or of "gaming" (Bevan & Hood, 2007). In other words, alternative logics are responses to interventions and as such they vary depending on the specific context in which the interventions take place.

In Pollit's perspective, alternative logics are both logics at the micro level, that is, of people whose performance is being measured, and the consequences of the clash of macro and micro logics. I believe that distinguishing these two elements of alternative logics, that is, the logic of a few people belonging to a group (micro level) and the consequences of the clash of macro and micro logics, enables us to achieve greater clarity when we lay out the second factor for the emergence of evaluation games. The consequences known as the *threshold effect, ratchet effect,* or *gaming* are not always the result of logics adopted by those who are subjected to performance management. Sometimes, the reason for which people start *gaming* is not because their alternative logic is different from that adopted by designers of the performance measurement system. It can be the case that these two logics are coherent but the situation in the workplace might change substantially thereby reframing power relations. Eventually, a new configuration of power relations might modify collective representations and the values shared by the community that is subject to the performance measurement.

In such cases, one of the above effects, like the *threshold* or *ratchet effects,* might be produced. However, these should not be conceived simply as micro logics pertaining to those in academia. They are, rather, diverse types of evaluation game, which involve not only players (those working in academia) but also specific rules and stakes. Moreover, the evaluation game often involves all parties implicated in power relations. For instance, the *threshold effect* is based on a rule about minimum targets: for example, the submission of three publications per researcher for every four-year period as part of the national evaluation exercise – as is the case in Poland (Korytkowski & Kulczycki, 2019). This target has the potential to motivate those researchers who produced only two publications in the previous period, but it could also demotivate those who produced nine publications. The latter might consider all their publications above the fourth as ineligible in terms of the evaluation exercise. However, this *threshold effect*, observed within a group of productive researchers, would not be produced only because of their adopted logics about "how to work" and "what good science is." The effect is better understood as an evaluation game caused by a redefinition of the context in which they work and the process of adaptation to it. Managers (i.e., a research organization in an institution) who use this particular regulation as the only stakes in the game are playing the game because they have the power to modify the stakes at their institutions by changing local stakes (i.e., increasing the number of publications which must be produced in a given period or introducing financial rewards for the best performers). Moreover, simply by virtue of the fact that they design and use the measures to govern and control, policy makers must also be viewed as involved in this game.

In further considering the three factors that give rise to evaluation games, one must answer two important questions. First, whether policy makers intend for these games to occur in academia and, second, which effects of research evaluation systems are intended as part of their implementation. In numerous studies on performance management systems, two terms, "effects" and "consequences," are used interchangeably. Moreover, the effects and consequences of performance management are mostly presented as intended or unintended. Merton (1936) analyzed the unanticipated consequences of purposive actions and highlighted that unforeseen consequences should not necessarily be identified with undesirable consequences. The consequences of an action are limited to the elements in the resulting situation, which are exclusively the outcome of the action. When researchers' daily work is analyzed, it is difficult to separate out the specific factors that determine changes in the way in which they practice science, that is, due exclusively to factor 1, factor 2, or factor 3. The problem of causal imputation in comprehensive social practices and social interventions should be a warning to us: It is very difficult to argue that particular new rules have changed practice. One should rather say that it is the implementation of those rules that has influenced the practice or co-contributed to the changes observed.

Following Merton's argumentation, while the unintended and anticipated outcomes of actions may be relatively (as regards other possible alternatives) desirable for the actor, at the same time, they may be viewed as negative, in value terms, to observers or subjects of the action. This is to say that when the state implements a new research evaluation system, it will entail certain intended and anticipated outcomes and effects. Those effects might however be perceived by those in academia as negative effects of this implementation.

Let us look more closely at the *threshold effect* described above, in which researchers must submit their best four publications every four years for the purposes of the national evaluation exercise. An intended and anticipated consequence was to motivate those researchers who were slightly below the threshold. But can one say that the demotivation of those researchers who were above the threshold was an unintended and unanticipated consequence of the implementation of this rule? One can say that it was an unintended but anticipated consequence. System designers knew that some researchers might be demotivated, but the primary aim of the system was to motivate those (directly or through their institutions) who did not perform very well, rather than to keep motivated those who were performing well. Nonetheless, an evaluation game (i.e., the *threshold effect*) could occur in such a situation. Can one say then that the evaluation game was an unintended consequence of the implementation of new rules, or is it the case instead that it was a foreseen and anticipated transactional cost of the social intervention? This example shows how difficult it is to determine whether something is an unintended consequence of a specific action, especially where a person is trying to associate unintended consequences with unanticipated effects. In light of this, it is fruitful to consider Lewis' idea (2015) regarding the explicit and implicit purposes of performance measurement. When policy makers implement a new research evaluation system, they present various explicit goals (e.g., improving research performance, motivating researchers to publishing in top-tier journals). An evaluation game, for example, the threshold effect, might be one of way of achieving the goals. Therefore, not all forms of evaluation game should be viewed as unintended and unforeseen consequences. Some might be intended or foreseen and in line with the explicit or implicit goals of the system makers. However, it is worth bearing in mind that the evaluation game is interactive among many actors, so one might justifiably ask why the intentions of policy makers should be privileged in assessing whether some effects are intended or no. I argue that their intentions are to some extend privileged because policy makers are real initiators and implementers of national or global research evaluation systems and as such play a key role in establishing the evaluation game. Moreover, drawing on the culturalist approach to science of Znaniecki (one of the founding father of science of science in the 1920s), the term "intention" does not mean here any mental state (of policy makers) but rather an action that is culturally meaningful and interpretable (Znaniecki, 1934). Hence, implementing a new evaluation regime can be understood as a communicative action, which involves policy makers (senders) and researchers and institutions (recipients). Thus, saying that some effect is intended means that some policy regulation has been received (implemented) and interpreted (influenced the practice of institutions or researchers) according to the goals of policy makers.

Dahler-Larsen (2014) argues that intentions themselves may not always be the best standard against which to assess the consequences of performance indicators. He therefore suggests that we use the concept of "constitutive effects" instead of unintended consequences, in order to be able to show that the use of indicators is truly political, because it defines categories that are collectively significant in a society. I agree with him that intentions are not the best standard and that differentiating effects based on whether they are intended or unintended is not only difficult to do but also does not provide us with useful tools for understanding what is actually going on with a particular social practice. Still, in discussions on the effects of research evaluation systems, policy makers and policy designers often claim that newly observed practices (which I would call types of evaluation game) were not intended. Hence, when they are under attack, policy makers often use the concept of unintended effects as a defensive argument. It follows that even if this distinction is not particularly useful from an analytical perspective, one should not totally give up on it precisely because it allows us to map policy makers' varied responses and reactions.

Merton (1936) drew attention to the fact that the consequences of an action do not only apply to those persons who are the target group of the action (and to the social structure, culture, civilization) but also to the actors themselves. In other words, the repercussions of a redefinition of the academic context extend both to academia and to those who use evaluative power to redefine it. Why then are actors not able to foresee all the consequences and to be prepared for them? Merton argues that during actions, actors ignore facts or make errors in their appraisal of them because they hold certain interests that blind them to risk or which create self-fulfilling prophecies. These mistakes in turn generate unintended consequences that can lead to other problems – through a chain of consequences – and to new solutions connected with new, unintended consequences.

The research evaluation system spans an extremely diverse reality, for which reason it must prioritize certain areas at the expense of others. This complexity can also pose an obstacle to the translation of findings from performance-management studies (based on the private sector) to the academia. Even the largest company will not contain as many different tribes and territories as the academia does (cf. Becher & Trowler, 2001).

Some researchers may claim that they do not play the evaluation game in academia, and are simply doing good research. However, evaluative power transforms the context in which they work, which means that they need to play in order to maintain the previous situation. In other words, it is not possible to be employed by a political institution of the state and not play; both the general situation and evaluative power itself force us to participate. Can people choose *how* they play, at least to some degree? Yes, I would argue that it is possible to a certain extent to design a strategy for playing. However, one needs to remember that the game is conducted differently depending on the context; thus it varies from central countries (like the Netherlands or the United Kingdom) which have dominant and privileged positions in terms of resources of all kinds to other countries in a lower position in the global distribution of power (like Botswana or Puerto Rico), and again, to aspiring countries (like Poland or the Czech Republic). The stakes in the game therefore also differ. I will return to this question in a later chapter where I investigate the diversity of evaluative powers. I turn now to an examination of the conditions and context that lay the grounds for the rise of research evaluation systems.

2

Economization and Metricization

Nineteen minutes before midnight on October 21, 2014, senior members of a medical faculty received an e-mail that started as follows: "If anyone is interested how Professors are treated at Imperial College: Here is my story." With these opening words, professor Stefan Grimm planned to inform his colleagues about the reasons for his decision. The e-mail, which he had intended to deliver late, was sent through a delay timer that had been set almost a month before its delivery. Professor Grimm was found dead on September 25, 2014 after he had been told that he was "struggling to fulfill the metrics" (Frood, 2015).

Stefan Grimm held a Chair in Toxicology at Imperial College London. For many years, he had maintained a successful publication track record that, in a publication-oriented academia, was a guarantee of stable employment. However in the past years, the nature of research work has changed significantly and British (and other) universities have become money-making research and education businesses. In this context, the number of publications is no longer deemed a sufficient measure of researchers' performance. Universities have also started to count the grant incomes generated by researchers, broken down by year. This metricization of academia – that is reducing every aspect of academic activity to metrics – has become all-encompassing.

Before his death, Grimm was told that any "significant funding" attributed to him had ended. Still, Grimm had tried to get more grants. Actually, as he wrote in his email, he was "the one professor on the whole campus who had submitted the highest number of grant applications" (Parr, 2014). On March 10, 2014, the head of Experimental Medicine at Imperial emailed Grimm and explained to him why he was struggling to fulfill the metrics: He needed to generate 200,000 pounds a year. In his last email, however, Grimm wrote that he was never informed about this and could not remember this requirement being part of his contract (Parr, 2014).

Although Grimm submitted the highest number of grant applications, the head of the division did not welcome this fact because, as he wrote, "grant committees can become fatigued from seeing a series of unsuccessful applications from the same applicant" (Parr, 2014). In today's academia, researchers must not only apply for external funding but above all continually be awarded grants. One can surmise that Grimm must have spent a lot of time writing proposals and reporting on his progress to meet "an acceptable standard" of performance. As a spokesman from Imperial commented, "it is standard practice in higher education institutions to conduct both informal and formal performance management" (Parr, 2014). As a standard practice, the monitoring and assessment of whether Grimm was performing at "the expected level for a Professor at Imperial College" was designed to ensure accountability in the use of research funds as well as profit increases for the researcher's employer. However, what the spokesman called a standard practice was not, according to Grimm, clearly communicated to him.

Grimm explained that researchers at Imperial College had to be granted a minimum amount of funding, but this amount was not stated in his contract. In this instance, by functioning in a veiled way, metrics and figures become an unspecified disciplinary power. Researchers are told that they have to meet certain funding levels, but these are never specified. This indeterminacy allows performance standards to be increased ad infinitum: When the threshold is not determined, it can be raised at any time.

Grimm argued that managers paid attention to researchers' performance only as expressed in numbers. He wrote that they "only look at figures to judge their colleagues, be it impact factors or grant income" (Parr, 2014). Research itself was not important, only publications ("impact factors") and money ("grant income") mattered. This is because these measures enable institutions to improve their performance in the research assessment of British universities. From the perspective of managers, this is a crucial goal of their work because it plays an important role in their career development.

Grimm's description of this system was candid: "These formidable leaders are playing an interesting game: they hire scientists from other countries to submit the work that they did abroad under completely different conditions for the Research Assessment that is supposed to gauge the performance of British universities" (Parr, 2014). According to Grimm's manager, Grimm was struggling to fulfill the metrics. Metrics, however, drive the entire system in which researchers, managers, and entire institutions are involved. More importantly, all parties involved in these power relations necessarily struggle to fulfill the metrics because there is no clearly defined goal to meet. When *profit margin* is a metric of an institution's performance and *publications* (as a sub-site of research) serve to increase margins, the metric fixation becomes a major driver of the science and higher education sectors.

* * *

In this chapter, I present a systematic account of two main forces within academic work – economization and metricization – which enabled the emergence of research evaluation systems. Economization promotes the idea that science's economic inputs and products should be utilized for bolstering the economy, while metricization reduces every aspect of science and social life to metrics. Therefore, the following account paints a broad picture that describes the social processes, values, logics, and technologies of power that ground today's research evaluation systems. With this perspective, it should be more than evident to research evaluation scholars and the academic community that metrics are just a symptom and not the cause of the problems in academia today.

In the Introduction, I argued that academia has become publication-oriented and that the whole research process is being reduced to publications. And yet during the past years, it has become apparent that producing good publications – that is, those which can be counted within research evaluation systems – is no longer enough in order to survive in academia. As Grimm argued, we are witnessing an interesting twist in the publish or perish research culture. According to him, "here it is 'publish and perish'" (Parr, 2014). In order not to perish, researchers must understand their work and its products not only through the prism of publications but also through its economic value as expressed primarily through the funding they have to secure for their institutions. This economic value relates not only to monetary value (the calculation of how much academic activities are worth in money terms) but also to a wider logic of profit. This is because researchers are impelled to think of the outputs of their activities (i.e., publications, grants) as commodities or resources that should be further accumulate. By obtaining grants, researchers partially relieve the university of the obligation of financing their research. In this way, they, rather than their universities become the financial managers of the scientific projects they undertake.

Stefan Grimm's death and the role of (in)formal research assessment in it might inspire critiques of all forms of evaluation and metrics. One can imagine two scenarios for changing the rules of metrics in academia. In the first, the academic community stops using any form of metric for the monitoring, reporting and management of institutions. Researchers should, however, consider such a scenario as a kind of thought experiment. This is because today's society, states, and institutions, both local and global, are products of modernization processes that have been rolled out through varied powers using such technologies as unification (of measuring systems, languages, religions) and counting (e.g., crops, subjects, or incomes). Because they are thus a key pillar of the social world, metrics cannot simply be dropped. In this view, metrics are only a technology of power, not power itself. An alternative perspective to this one understands metrics as holding a degree of power in themselves (Beer, 2016). If one looks at the history of universities, one can see that they have always been implicated in power relations with those who founded and financed them (Axtell, 2016). As a consequence, if it was not metrics that were being used to manage them, it was religious or state imperatives. In other words, diverse technologies have always existed (metrics, funding levels, or even the military) that have served to maintain power relations.

In the second scenario, the academic community proposes the design of improved metrics and more responsible ways of using them. However, even where it puts forward better metrics, any critique focused solely on metrics' abuses (and their tragic consequences as in the case of Stefan Grimm's death), can be no more than a process of faultfinding incapable of offering up constructive solutions. Social scientists have known for decades that metrics and evaluation regimes would change social practices (Campbell, 1979). Thus, to suggest that one design better metrics – understood as metrics better suited to the "objects" they measure – and use them only in appropriate and responsible ways (Wilsdon et al., 2015) cannot substantially improve the situation in academia.

Even if one were to formulate better metrics for scholarly communication, the negative or unintended consequences of metrics fixation, such as exploitation and burnout in academia, would endure. For instance, over the past few years, it has become possible to analyze the impact of individual articles instead of whole journals thanks to, among others, so-called alternative metrics (Bornmann & Haunschild, 2018). However, this solution has not changed the fact that journal impact factors are used for assessing the productivity of individual researchers. Instead, now both journal- and article-level metrics are used to assess institutions and researchers. Thus, the core problem is neither metrics like the amount of external funding generated by a researcher annually nor the number of publications published by that researcher. The problem is that research and its results are monitored, reported, and assessed through the prism of economic value. And yet the process of subordinating academia to power does not end at the reduction of scholarly communication to publications. Rather, the process is completed when the whole of the higher education and science sectors are reduced to inputs used in the production of economic value. This process implies that various noneconomic forms of capital (e.g., social, cultural) possessed by research institutions and universities are also converted into economic capital. However, knowledge, as a key basis for the activities of these two sectors, is by its very nature not completely commodifiable. This kind of total commodification would require reducing knowledge to inputs in the production of value. Not all academic activities can be directly translated into economic value and the translation of some activities is a complex process (Jessop, 2017; Reitz, 2017). Many aspects of such translations are mediated, among others, by the economy of prestige, where prestige is understood as a value based not only on economic but also sociocultural capital, or on the need for recognition and the distribution of credit in science (Merton, 1973).

Assuming that neither of these scenarios can solve the salient problems in today's academia, what can be done? If researchers criticize metrics because they consider them to exert undue pressure on scholars, then in their search for solutions, they must look beyond metrics. In seeking to ameliorate conditions in academia, they must therefore concede that metrics sustain social (global) processes and are simply tools that are used for the realization of certain values. On the following pages of this chapter, I describe how two main forces, economization and metricization – which I define as logics of social processes – produce varied pressures on academia and lay the grounds for the rise of research evaluation systems. Clarifying this issue requires that we show how these logics are associated with different values and elucidate the impact this has on the design and use of technologies of power such as evaluation regimes.

2.1 The Logics of Social Processes

Discussing the Stefan Grimm case and the (in)formal evaluation policy of the institution that employed him, one can see that the origins of the compulsion to produce ever more publications and to increase profit margins are external to university institutions. One must understand these pressures as part of a global movement that is internalized by individual scholars who self-discipline in order to meet the expectations of evaluation regimes. In the case of the institution that employed Stefan Grimm, the force was exerted by the British government through its introduction of the Research Excellence Framework (REF) and competitive sources for research funding. It is not only the state, however, which is only one among numerous actors at the global level, that produces this kind of pressure. Other forces like international publishing houses or global bibliometric databases are also behind it. Further, all powerful actors involved in research evaluation are influenced by global historic and present-day processes and their logics that constitute the context that gives rise to research evaluation systems. By examining this context, we can understand relations across social processes like bureaucratization and neoliberalism, the logics of processes like economization and metricization, and their attendant technologies of power, such as university rankings or the central planning of science.

I distinguish three intertwined dimensions of this context: social processes, logics, and technologies of power. Doing so enables me to show why any reduction in the negative consequences of the use of metrics in academia cannot be based only on a critique of metrics but must aim rather at transforming academia's fundamental values. With respect to the distinctions that I draw between dimensions, it should be noted that while it is possible to maintain these at the analytical level, where each dimension is mutually distinguishable from the other, in practice, the lines are less firm. Thus when a specific case is discussed, the boundaries between and within dimensions begin to blur (e.g., is cooperation a social process or rather a value?). This is due both to the mutual intertwining of these dimensions and to the specificity of concepts such as value, social process, and logic. Any exposition of a socio-historical context has to confront such challenges as they are inherent to the process of conceptualization itself. However, limitations notwithstanding, such conceptual tools can still be valuable for deepening our understanding of the context.

The first dimension of this context relates to social processes like bureaucratization (Weber, 1978), neoliberalism (Harvey, 2005), quantification (Espeland & Stevens, 2008), the rationalization of society (Vera, 2008), and scientific revolution (Bernal, 1971). These processes constitute modernization and are grounded in values like competition, cooperation, development, freedom, profit, and well-being. Social processes are manifested in the practices of various institutions and their members who determine their own institutional aims on the basis of accepted values. My definition of values draws on Znaniecki's approach, for whom values are cultural objects that shape human activity in essential and practical ways (Znaniecki, 1934). In relation to a particular system of values, a value may appear as "desirable," "undesirable," "useful," or "harmful" in connection with the other values. For instance, I therefore treat competition and cooperation – which might be both desirable or undesirable depending on the context and actors involved in the action – as value grounding social processes and not processes that are triggered by specific values. Any attempt to identify a comprehensive set of historical and dynamic processes and values must be considered suspect and the distinction between values and processes might also be questioned. Nevertheless, I believe that it is both possible and fruitful to trace the key modernization processes that transformed academia in a global context and allowed research evaluation systems to arise.

The social processes identified in the first dimension can be linked to two overarching concepts. The first is modernity which, as a particular set of sociocultural norms, has been produced, among other things, by such processes as the rationalization of society, scientific revolution, and bureaucratization. Modernity is not only a specific historic period that can be defined within the bounds of geographical space; it is also a way of conceptualizing the rise of (modern) nation-states. The other overarching concept is capitalism, conceived of as a type of economic system whose origins are intertwined with modernity. Depending on whether primacy is given to modernization or capitalism, it can be argued that capitalism was a precondition for the emergence of modern states or, on the contrary, that it was the development of modern states that enabled capitalism to emerge. Capitalism has changed shape many times since then, and one of these transformations led in the 1970s to the emergence of neoliberalism (Bacevic, 2019; Harvey, 2005). With respect to science, neoliberalism has played an important role, transforming the science and higher education systems by embedding market logics within them.

The second dimension of the context comprises the logics (or forces) of social processes, that is the *modus operandi* of a given process. Logics, which in this instance refer to regulatory processes, are the sets of procedures and principles pertaining to the processes identified in the first dimension. Therefore, logics are connections between values and processes. For instance, with regard to neoliberalism as a social process, one can indicate such logics, among others, as economization (Berman, 2014) or accountability (Baldwin, 2018); for the rationalization of society, the relevant logics might include economization and metricization (Muller, 2018). A single logic can drive various processes; for example, economization might serve as either the logic of neoliberalism or that of rationalization. From the perspective of the three-dimensional approach I am outlining, academic capitalism (Slaughter & Leslie, 1997), accountability (Watermeyer, 2019), the knowledge-based economy (Sum & Jessop, 2013), New Public Management (Dunleavy & Hood, 1994), and publish or perish research culture (Moosa, 2018) are all logics of grand social processes that we can observe in academia today.

The third dimension comprises those tools that serve the running and maintenance of logics. For instance, research evaluation systems (Hicks, 2012), productivity indicators (Godin, 2009), university rankings (Hazelkorn, 2015), peer review processes (Moxham & Fyfe, 2018), and central planning (Hare, 1991) are "technologies of power" through which logics are implemented and operate. These technologies are shaped according to the logics and values on which they are grounded. For instance, research evaluation systems and metrics are tools that serve to reproduce various regulatory processes like economization or accountability. Important here is the fact that the form these technologies of power assume depends on the logics governing a given social process. These logics are in turn adjusted to processes on the basis of grounding values.

Let us now consider how these three dimensions of a context – social processes, logics, and technologies of power – work, and outline the resultant implications for an analysis of the effects of research evaluation systems. For example, one can ask what a process such as bureaucratization might look like in a given institution (e.g., a university) depending on which dominant values ground the process. Understanding bureaucratization as the establishment of an administrative system in a university that serves to govern and manage that institution, I analyze it on the basis of two values: competitiveness and the common good.

With respect to the first, bureaucratization can be understood as a solution that supports competitiveness both between academic staff members at a university and between the institution and other universities in the global higher education market. If a university administration is instructed that global competitiveness is crucial for the institution, then various indicators and metrics serving to measure both academics' internal productivity as well as that institution's place in the market can be used. In this way, bureaucratization grounded on competitiveness introduces metricization as its own logic. Hence, various technologies of power like bibliometric indicators or university rankings are used to stimulate successful competition with other institutions and competition among researchers from the institution with each other, in order that these ultimately lead to a better position in the rankings. It is very likely that this logic of metricization is combined with other logics like economization and academic capitalism, which create a monetary reward system and pay for publications in top-tier journals that are counted in the university rankings. While these strategies might improve the institution's ranking, at the same time, they might damage internal cooperation within it.

Bureaucratization could, however, be based on a different dominant value, for instance, the common good, which is my second example. The role of an administrative system in such an institution is to contribute to increasing the quality of education that would serve to reproduce the knowledge-based economy. Various metrics might be used to measure academics' effectiveness (for instance, the research-based learning courses number) or their engagement in the university's third mission (Rosli & Rossi, 2016). If the success of an institution is determined by its capacity to nurture a generation of more effective teachers and not by its position in the university rankings, an institution would be keen to promote cooperation between its staff so that together they might achieve its goals. In such a case, the administrative system would not introduce pay for publications in top-tier journals because that technology of power would not support the maintenance of its core logic.

In the first example, the university could also promote cooperation along with competitiveness, designing different metrics to both incentivize competition with other institutions and cooperation among its own staff. This is the crux of the relationship between adopted technologies of power and logics based on different values. Depending on the values that set overall objectives, different technologies are used, or the same technologies are used with varying consequences in different institutions. In other words, the measurement of publications can have diverse effects, but the effects of using such metrics must be evaluated in the broader context and in connection with the other technologies, logics, and values that drive a given institution. Thus counting the number of publications can both discourage cooperation (when a university, according to its monetary reward system, does not pay for co-authorship publications with authors from the same institution) and encourage it (when a university pays the same amount of money for multi- and single-authorship publications). In relation to this point, it should be stressed, however, that institutions are driven by varied, sometimes conflicting, values, and aims. Monetary reward systems can therefore be gamed. For instance, when a university pays for multi-authored publications, it might incentivize artificial cooperation within the institution which might produce an unintended ("this is not actual cooperation but an incentive driven practice") or an intended effect ("we want to increase the level of cooperation by all means") of this technology of power. It is the grounding value of a given logic (e.g., competitiveness or cooperation) that allows an administration to assess the effects of a proposed solution, in this case, the introduction of a monetary reward system.

Furthermore, it should also be noted that universities are institutions of the longue durée (Braudel, 1958) and over the course of the many years of their existence, different logics and sets of values are adopted. Although one can certainly impose a completely new logic or set of values on an institution, the legacy of previous logics, rooted in different procedures and practices, will affect the functioning of the new logic and may modify or distort it. This can be seen, for example, in those universities which for many decades were primarily geared toward teaching, and which in recent years have had national scientific policies promoting an intensification of their research activities imposed on them (Antonowicz et al., 2017). The implementation of new technologies of power and logics always has to confront existing logics that constitute the historical and cultural context in which such implementation takes place.

Above I have highlighted the fact that these three dimensions are intertwined and can be distinguished only analytically as elements of a whole context. It is not possible, moreover, to identify all phenomena operating in each of the dimensions. One can only focus on those processes, logics, and technologies of power that have a substantial explanatory power within the adopted perspective. Distinguishing three intertwined dimensions is the first step in investigating the context in which research evaluation systems arise. The next step is to show how historical processes – both major ones like bureaucratization and minor ones like the implementation of research evaluation assessment in a given country – can be investigated and used to reconstruct the context of today's technologies of power.

It is because technologies of power are mediated by (historical) processes and through varied logics that one can develop a reconstruction of the broader context through an account of the dimensions described above. For instance, any investigation of the REF in the UK, that views it as a technology of power and shows how this science policy instrument emerged thanks to, among others, such logics as economization and New Public Management, must also clarify the assumptions through which this instrument works, in other words, the values it is grounded on. The values to which the REF refers are presented (1) *implicitly*: in no document can one find statements related to the objectivity of numbers as a value grounding a science policy instrument. However, through an in-depth reading of policy papers, such a value can be analytically derived; and (2) *explicitly*: It is possible to find references to such values as global competitiveness in science, or the profits that can be generated for a given country by the development of universities.

When investigating current research evaluation systems, scholars have to keep in mind the fact that social processes are, by their very nature, dynamic. Thus they can best be apprehended through their logics, that is, their ways of functioning. One can say that in these logics, social processes both solidify and objectify themselves, thereby becoming potential research subjects. Moreover, through the objectification of procedures, rules, and regulations, global processes begin to be deployed at lower levels, that is at the national and local levels. Logics start to be adapted not to the whole social world, but rather to more narrowly defined parts of it. Distortions of these logics are caused by different sociocultural contexts and, in the case of the public sector, by policy aims and instruments.

In Chapter 1, I described the practice of publishing *Festschrifts* and argued that researchers often have to adopt a dual loyalty because of the conflict between the values that drive their actions (loyalty to a "global" discipline versus loyalty to a "local" institution). Such tensions, however, can be produced not only by a conflict of values (or loyalties) but also by tensions between distinct versions of the same logic. Social processes and their logics are global (e.g., neoliberalism and economization), and frequently, so too are key actors (e.g., the OECD), World Bank, Elsevier, and Clarivate Analytics). However, when logics begin to be adapted to the context of a particular country or institution, tensions between different versions of them can occur. For instance, some institutions might acknowledge only journal articles published in top-tier journals (according to the publish or perish logic at the global level), whereas other universities – in the context of national science policy – might acknowledge all publications indexed in an international database like the WoS. Researchers may recognize the same logic at play, but each of their publications – if for instance, they are considering research mobility or cooperation – have to be taken into account separately in light of each of these versions of the publish or perish logic. Apprehending the importance of national and local versions of global logics is crucial for understanding why the same metrics can produce variable effects: "one (countable) publication" will not be understood in the same way in all places.

In this chapter, I outline two main logics ("main forces") about academia and academic labor at the turn of the twentieth and twenty-first centuries that are key pillars of the context in which the rise of research evaluation systems took place. These forces, although distinguishable, are two sides of the same coin and do not exist without each other. The first force is *economization*. Its key elements can also be found in other logics like academic capitalism (Slaughter & Leslie, 1997), the knowledge-based economy (Sum & Jessop, 2013), and New Public Management (Dunleavy & Hood, 1994). Within the perspective presented in this book, economization is a logic (or force) of neoliberalism. The other force is that of *metricization* – a logic of quantification – of academia and academic labor. This is combined with the publish or perish research culture (Nabout et al., 2014), the concept of metric-wiseness (Rousseau & Rousseau, 2017), audit culture and society (Power, 1999; Strathern, 1997), various indicator games (Bal, 2017) or gaming, the phenomenon of university rankings (Kehm, 2014) and their varied critiques (Kehm, 2014; Pusser & Marginson, 2013), and quantifying the (academic) self (Moore, 2017) as well as with broader concepts such as trust in numbers (Porter, 1995) and central planning "for" and "of" science (Graham, 1964). These forces are played out differently at three mutually linked levels: global, national, and local. The "glonacal agency heuristic" introduced by Marginson and Rhoades (2002), which I adopt, allows me to highlight the simultaneity of flows of logics across all three levels. Thus while metricization or economization might first occur in the higher education or science sectors at the global or institutional levels, eventually, these logics also influence contexts at other levels. Today's academia is neither local, national, nor global. It is global, national, and local all at once. The tragic case of Stefan Grimm illustrates the fact that academia and universities are strongly focused on profit margins and economic development, which are consequences of neoliberal shifts in academia.

2.2 The Economization of Academic Labor

Academic labor and the academic profession are in a process of constant restructuring. In the second decade of the twenty-first century, researchers tend to believe that such working conditions as we currently have, focused on productivity and accountability, never existed in the past. This is a partial truth, akin to believing that at the beginning of the twentieth century, universities around the world were organized according to Humboldt's ideals. The pressure on academic work to be systematic, rigorous, and reportable is much older than the emergence of New Public Management with which, in debates on the subject, these phenomena are often linked. Over the centuries, it is certainly the case that the content of this pressure has varied, which is largely due to changing types of contract between states and universities. The bureaucratization and rationalization of society, as Weber (Vera, 2008; Weber, 1978) shows, render all social actions disciplined and methodical; thus academic labor, as well as research or scientific investigation, also have to be systematic and disciplined. The monitoring, reporting, and evaluation of reported actions is one of the key expressions of a rationalized society that is sustained through various forms of bureaucratic action.

In academia today, researchers often stress the fact that they have to be more productive than their colleagues had to be 30 or 40 years ago, in order to achieve similar positions and stable working conditions. Peter Higgs, whom I quoted earlier and who is a Nobel Prize laureate in physics, commented that he would not get employed in today's academia because he would not be considered "productive" enough. In the era of academic capitalism (Slaughter & Leslie, 1997), stable labor conditions are a luxury for which researchers have to pay ever more. This price, expressed by the degree of productivity (e.g., the number of papers or external funds) is getting higher while at the same time the availability of the commodity in question (e.g., tenured positions) is shrinking. Researchers have to produce more papers than their colleagues before them did, and yet even so, there is no guarantee that they will obtain the same conditions as their colleagues enjoyed decades ago. Being a researcher in the current moment means accepting (or at least respecting) a type of work which is measured for at every step (Watermeyer, 2019). Today's academia and academic labor are sometimes perceived through their acceleration (Vostal, 2016), which is manifested through various (self) quantification practices (Angermuller & Van Leeuwen, 2019; Davis et al., 2012; Espeland & Stevens, 2008) and through the translation of every piece of academic work into the language of economic values (Berman, 2014; Münch, 2013). Moreover, the salient economic narratives of the last four decades, based on the concepts of globalization (King et al., 2011), competitiveness (Sum & Jessop, 2013) and the knowledge-based economy (Jessop, 2005), make it evident that research and science are interconnected in a global network of power relations, sociocultural, and economic capital.

Neoliberalism, in the perspective elaborated in this book, is what brings together economization and metricization. It is based on a set of key principles through which society as well as its culture, economy, and organization are understood and governed. Olssen and Peters (2005) argue that four principles or rather presuppositions can be identified in relation to it. First, within neoliberalism's framework, individuals are economically self-interested subjects. Second, the free market is the best way to allocate resources. Third, the free market regulates itself better than any other outside force, and fourth, trade must be free. These principles started to also be implemented within the public sector in which relations between government and society had already been transformed through New Public Management's ideas. This is a paradox, since the introduction of these principles to the public sector meant going against the nature of the public sector: After all, markets in the public sector are strongly regulated. Neoliberalism is a form of political–economic organization, and it is a political project that serves to restore the power of economic elites (Harvey, 2005). This social process is carried out in part, through the financialization of every aspect of life; it implies a shift away from production (of goods, knowledge, etc.) to the world of finance where capital may be accumulated without having to produce anything. Here it is worth emphasizing the ideological aspects of this process, that is, the maintenance and expansion of the sphere of influence of certain values like individuality, freedom of choice, and laissez faire.

There is in fact a contradiction here, because, at least for a critical political economy, it is impossible to claim that capital is being released from production.

In higher education and science, the neoliberal shift has introduced a new mode of monitoring, reporting, and evaluating that has transformed the way in which academia is governed and reproduced. In the late 1990s, scholars started to use the label "academic capitalism" to name this new context (Slaughter & Leslie, 1997) in which new forms of competition arose in academia. A decade earlier, the knowledge-based economy had been one of the master narratives guiding the transition to post-Fordism, a system of economic production which relied more heavily on information, communication technologies, and flexible and digital labor (Jessop, 2005; Musselin, 2018). This narrative first appeared as a theoretical paradigm in the 1960s, but it was in the 1980s to 1990s that it became a policy paradigm oriented to competitiveness that was promoted by the OECD.

Certain key tenets of neoliberalism also make a good fit with the characteristics of academic work. Thus the proliferation of academic capitalism and the knowledge-based economy in various local and national higher education and science contexts was an inevitable result of the expansion of the neoliberal paradigm. Marginson and Rhoades (2002) show that the neoliberal shift in higher education manifests itself through a reduction of state subsidization of higher education, the shifting of costs to the market and consumers, the demand for performance accountability and the emphasis on higher education's role in the economy. This economic role had to be promoted in the new era of the knowledge-based economy because during the crisis of Fordism, education was criticized for failing to meet the needs of the labor market's changing economy (Sum & Jessop, 2013). In this situation, academia as an environment in which higher education and research are produced, has had to prove its capacity for competitiveness within the knowledge-based economy's global market.

Today science is no longer perceived of as either a community of scholars or as the republic of science. It has become part of a globalized economy and the key actor in the production of knowledge and its transfer to governments, industry, and the public. Within a neoliberal approach, universities and research institutions have to compete for resources (funds, people, and infrastructure) to produce the best and most useful knowledge in the shortest time possible. The economization of this sector is mediated by a form of prestige economy (Blackmore & Kandiko, 2011). Without this element, the translation of changes to the research institutions themselves from changes occurring at the level of the global economy would be fragmented. It is not the case that neoliberalism, on penetrating the academic sector, revolutionizes it completely. Rather, the process encounters the sector's own specificities (especially prestige as a flywheel for competition in science), which it transforms, but does not exchange for completely different ones. Thus the traditional model of university governance, linked mostly with Humboldt's ideals, has had to be transformed and replaced by tools that enable competence in the global race for resources in the world of knowledge production.

Berman (2014) contends that if neoliberalism is perceived as the idea that it is desirable that free markets organize human activity and that governments should focus on securing property rights and promoting free trade, then the transformation of science over the last decades should be interpreted through the prism of neoliberalism. According to Berman, this transformation manifests itself in "the expansion of intellectual property rights, the idealization of entrepreneurship, and the reorientation of academic science toward work with commercial value" (2014, p. 398). Berman's research, however, demonstrates something still more important: that in analyzing the transformation of the scientific landscape, we – as higher education and science scholars – should not limit our analyses to neoliberalism, evaluating all developments through the idea of free market fixation. This is because in doing so, we might miss the broader trend in science and technology policy, which Berman terms economization, and which is defined through two phenomena. First, economization involves a trend toward thinking in terms of the economy by increasing political concern with "the economy" and economic abstractions like growth or productivity as an objective that governments can pursue through public policy. Second, economization consists in the definition of ever more types of activities as inputs that can be incorporated within the system of science. As a result of these two phenomena, Berman shows that governments understand their role as to positively influence the economy because they assume that the relation between science, technology, and the economy is clear and that an understanding of this should inform policy. While for Berman, economization is a process that is broader than neoliberalism, in the perspective that I elaborate in this book, economization is a logic (or a force) of neoliberalism itself. This difference in analytic approach, however, does not alter our understanding of economization and its consequences for modern science.

One of the main outcomes of the economization of science is the understanding that research's core goal should be to provide the desired economic results. Thus, during the 1980s and 1990s, intellectual property rights and entrepreneurship became key buzzwords for academic management, replacing the objectives of solving scientific and technological problems. Economization has distorted the republic about which Polanyi (1962) was writing. In times of the economization of science, researchers can freely choose scientific problems to investigate, but only when the results of such studies provide the desired economic results and positively contribute to the economy or, at a smaller scale, generate an acceptable profit margin for research institutions.

It is of course true that science has never been a pure republic in Polanyi's sense. There have always been certain areas, such as the military or public health, on which the state has focused its attention and defined expectations and incentives. This fact is not specific to recent decades and can be observed in much earlier periods. For instance, the first American colleges of the eighteenth and nineteenth centuries operated as public–private undertakings that served provincial legislatures and had to have a positive impact on local economies (Axtell, 2016). According to Berman (2014), however, prior to the 1960s, such areas were not identified with the (national or global) economy in the United States. Nonetheless, as I argue below, in an overlooked part of the world, the Global East, such areas were identified with the economy much earlier and science was perceived as a direct tool for shaping the national economy. In the Russian and Soviet contexts, during the 1920 and 1930s, science and technology were understood as tools of modernization (e.g. through electrification) and as key elements of a centrally governed economy (Dobrov, 1969b). Thus, one can enrich our analyses of the consequences of economization by looking at the history of the socialist economization of science.

In periods of economization, science resembles Bernal's ideals of it much more than Polanyi's. As laid out in Chapter 1, the primary function of Soviet science was not only to solve scientific puzzles but also to serve soviet society and its economy. Every scientific activity and technology enterprise conducted by an institution had to be assessed in light of its capacity to create the desired economic results. Certainly, in the case of Soviet science, those activities were centrally planned, whereas in the current era of the economization of science, the free market is viewed as a ground for scientific activities that can contribute to the economy. Nonetheless, despite these significant differences, both approaches can be said to have substantially transformed higher education and science in all parts of the world.

It is economization that led to the fact that researchers like Stefan Grimm from Imperial College London started to be assessed through the "significant funding" they provided their institutions, while these institutions began to be evaluated on the basis of the level of their productivity as expressed by profit margins, the number of top-tier papers, or their position in various university rankings. At each level of academia, reporting and monitoring, which are well known features of any bureaucratized society, have been combined with accountability practices in relation to public or private funds provided to universities. Science and researchers then became a part of a global knowledge-based economy, and research institutions and researchers started to be understood, monitored, and assessed as producers of knowledge. And as in other areas, they started to be measured for through various numbers, thresholds, and expectations. In other words, the implementation of the logics of economization caused, at the same time as they also required, massive metricization of the whole science and higher education sectors. Without measures, metrics and numbers, academia could not be financialized, that is pushed into the logic of economization.

2.3 Metricization as Reduction

The propensity to measure has a long and complex history (Desrosières, 2002, 2013; Kula, 1986). Those aspects of it that are especially relevant to this book developed in the seventeenth century when the emergence of a systematic science of social numbers was used to create social policy and to describe society in terms of various metrics. This practice was based on the trust in numbers that was produced by the culture of objectivity built by modern science (Porter, 1995). For instance, William Petty argued that quantitative knowledge was crucial for England's colonization of Ireland and for public finances, while his friend John Gaunt analyzed weekly mortality statistics for London to create life tables that estimated the probability of survival for each age (Mennicken & Espeland, 2019).

However, it was only in the nineteenth century that statistics came into common use. In the first half of that century, the idea of the *average man* was born, and this was one of the key milestones in the development of statistics for public policy purposes. As Desrosières (2002) shows, this notion was based on the generality of Gaussian probability distribution and on sets of "moral statistics" developed by bureaus of statistics, which formed the basis for perceiving statistics as powerful tools of objectification. Since then, calculations based on numbers have provided the solid foundations on which management of the social world is based. Statistics allow a large number of phenomena to be recorded and summarized according to standard norms. Thus reporting-based bureaucratization has gained a powerful ally. Phenomena were not only described by officials and written down in standardized forms, they were also ranked and could therefore be easily evaluated, which practices laid the ground for further scientometrics (Godin, 2006).

Before explaining what the metricization of academia is, distinctions between key concepts must be made because in the literature, varied terms like *measure*, *measurement*, *metrics*, *indicators*, and *quantification* are used interchangeably. Moreover, each of these concepts has diverging meanings across fields and science sectors (Mari et al., 2019; Wilson et al., 2019).

The basic concept at play here is *measure*, that is a unit used for stating the property of an object, phenomenon, or event such as its size, weight, or importance. Thus, measure can relate to qualitative or quantitative properties. For instance, Kula shows how a bushel was used as a measure of grain. *Measurement* is the assignment of a number to a selected property. When it is defined in this way, measurement is understood as *quantification,* that is as a mapping of the results of the measuring of quantitative properties onto numbers. Using the example of bushels, one can say that X person harvested 10 bushels during one day (day being a measure of time). Measurement is different from expert assessment because the results of measuring do not depend on who is doing it, whereas expert assessment depends not only on rules and assessed objects but also on experts. This important characteristic of measurement is perceived as the foundation of its impersonal nature and objectivity.

Quantification can also be understood as a setting of the conventions on how to measure and interpret the result of measuring. Kula (1986) shows that early measures had a substantive character, which means that they represented something in relation to man's personality or the conditions of his existence. Thus, the size of a field was determined by the amount of human work (in terms of time) needed, for instance, to sow it with seeds. Modern and universal measures like the square kilometer, however, have meaning only through their acceptance within given community conventions. These earlier measures were also meaningful only on the basis of conventions adopted by a given community; the difference lies in the abstract character of new conventions. As Power (2004) argues, such conventions become naturalized and taken for granted by the communities of practice in which they have relevance. In this way, the institutional and conventional character of measures becomes invisible, and measuring emerges as common sense.

Measure has no inherent meaning. Its meaning is instead constructed through accepted conventions on how to measure and how to interpret measure's results. In this way, quantification draws attention to the social nature of setting measures and rules of measuring and of mapping the results into numbers. Seeing something through measuring and quantifying it is the first step to controlling it. The social character of quantification is institutionalized and legitimated by implementing it in various public and political contexts. Hence, this conventional process has actual and material effects and also influences the measured areas that process sociologists of quantification term "reactivity of measurement" (Espeland & Sauder, 2007).

Metric is a measure that contains rules and norms as to how to carry out measurement and how to interpret the results of measuring, that is how to construct the meaning of the measure in relation to other comparable measurements. One can therefore say that a metric is simply a quantification. However, metrics are often derived from two or more measures and as such could be understood as aggregations of several quantifications. Nonetheless, metric and quantification are often used interchangeably. If the meaning of measure is the result of a convention accepted by a given community, then a change in social and power relations can determine a change of convention and, as a result, redefine the measure.

Thus those who have power can also set measures and conventions or, in other terms, use metrics. Returning to the example used in an earlier chapter of a bushel as a measure, when the merchant was selling the bushel of grain, the bushel would be stricken or leveled. Yet when someone repaid the grain to this merchant, the bushel needed to be heaped or "topped up" because the merchant had the power to enforce it (Kula, 1986). In this way, in practice, using the bushel involved actually using a metric because in a given community of practice, a particular measure (the bushel) was associated with an accepted (or at least respected) convention on how to carry out measurement.

Indicator refers to a collection of rank-ordered (or aggregated into an index) data. Often in the form of numerical data, indicators represent the performance of specific units like objects, phenomena, or events and also serve to map results of measurements into numbers. However, not all numerical data are indicators. As Davis et al. (2012) show, the differences lie in how indicators simplify "raw" data (by filtering, replacing some data with statistics such as means, filling missing data in with estimated values, etc.) and name the product of this simplification. This process of naming is just as important as simplification. By calling an indicator a measure of the "development of something," the producers of this indicator assert that such a phenomenon exists and that their numerical representation measures it. In this way, as Rose (1991) shows, public order, for example, manifests itself through the crime rate and the divorce rate becomes a sign of the state of private morality and family life.

In the case of academia, the history of one of the most important indicators, that is the productivity of institutions or researchers, teaches us how understanding of the measured phenomenon by a given indicator can evolve (Godin, 2009). Initially, productivity applied to science was understood as *reproduction* (of men of science), and then as *output*. Later, thanks to, among others, the role of global actors like the OECD), understanding of productivity in the field of science attached to the concept of *efficiency*, which was later understood in the style of neoclassical economists as *outcome*.

Now, through the example of measurements and indicators used in the academia, I will look at how these distinctions work in practice. Taking into account that the process of naming an indicator is important and serves to constitute the existence of a given phenomenon, let us start from one of the biggest buzzwords used in academia, "research excellence" (Antonowicz et al., 2017; Münch, 2013). As a representative example of how to define and measure this concept, let us take the report *Innovation Union Competitiveness report: Commission Staff Working Document* presented by the European Commission (Directorate-General for Research and Innovation, 2014), in which research excellence is defined in the following way: "Excellence in research is about top-end quality outcomes from creative work performed systematically, which increase stocks of knowledge and use it to devise new applications" (p. 108). To measure research excellence, a composite indicator, consisting of four variables (metrics), is introduced: highly cited publications, top world-class universities and research institutes, patents, and the value of European Research Council grants.

Investigating this definition of research excellence, one easily finds that "research excellence" cannot exist outside rank-ordered data because it is defined as "top-end quality outcomes." Thus, the research excellence of some countries, institutions and so on can be ordered and an indicator used to rank the results of measuring certain dimensions of the concept. In this approach, these dimensions exist only when they are defined through measures that can be used to grasp them. Thus, important questions arise: Who chooses the measure and why are all metrics based on publications and external funding? Who selects the databases that perform the measurements? Who measures? Who sets the conventions on how to measure and how to interpret the results of measuring? Providing the answers to these questions is only partially possible. Measuring science and research evaluation is a black box: at the outset we have science policy objectives (like a more strategic focus on world competition), whereas the output consists in a set of measures. What we – as the academic community whose work is measured – cannot see is how various political and policy interests shape the choice of measure and the setting of conventions for using metrics. In the European Commission's document, research excellence, a very comprehensive and broad concept, has been simplified and captured in four metrics. Only then can various countries and entities be easily compared in the sense of objective assessment. This reduction of every aspect of social life and the world to metrics is what I term metricization.

The metricization of contemporary academia is rooted in the demand for accountability for the performances of institutions and individuals. Metrics, as Beer (2016) notes, have had an ordering role in the social world for a very long time. Accountability includes varied practices but making institutions accountable usually means making them "auditable," which involves devising metrics or indicators to measure performance (Angermuller & van Leeuwen, 2019; Espeland & Sauder, 2007). Power (2004) argues that auditing is the administrative equivalent of scientific replication understood as the institutionalized revisiting of performance measures. Interestingly, the penetration of audit culture into science and higher education sectors has been enabled not only by the spread of the New Public Management but also by the revolutions that have taken place within the education sector itself. Strathern (1997) describes the pedagogical revolution of the second half of the eighteenth century related to the introduction of the examination of students. The introduction of the idea that if you can measure human performance, you can also set goals and measure improvement. Although auditing is most often associated with businesses, the combination of human and financial performance occurred not in business but the field of education through writing, grading, and examining practices. Strathern emphasizes that the development of measurement technology in education has contributed to the development of standardized auditing procedures that are now routinely used in financial auditing in universities. As Strathern (1997) emphasizes: "Academic institutions were re-invented as financial bodies" (p. 309).

While Michael Power was writing the preface to the paperback edition of *The Audit Society* published in 1999, he was serving as the head of a department at the London School of Economics and preparing his department for the Research Assessment Exercise in 2001. Power, while well aware of the role that auditing plays in science, points out that during this preparation of British universities stories were circulating in the academic community both about the absurd consequences of auditing and "strategic necessity of 'playing the game'"(Power, 1999, p. xv). This is because the (evaluation) game is inevitable and the stakes are important for players in the higher education and science systems. Auditing, like research evaluation, is recognized as a widespread phenomenon that has been influenced by the New Public Management in today's society and in science and higher education sectors (Shore & Wright, 2003; Strathern, 1997). Nonetheless, audit and evaluation are distinct in important ways. One of the key objectives of audits is detecting frauds, whereas the mechanism of evaluation is not to find errors or fraud, but rather to control and shape social practices. Although auditing is also credited with changing reality as one of its key objectives, the specifics of the two technologies are different. Audit checks for errors, assesses, and verifies the compliance of an audited practice with predefined rules, processes, and objectives. Evaluation, on the other hand, serves to determine value and as such has diagnostic utility and is intended to show potential ways of the development. However, Strathern might disagree with this distinction, because for her auditing and measurement itself are combined in the concept of "measuring the improvement." Nevertheless, it can be considered that both evaluation and audit are manifestations of the broader phenomenon of using metrics to manage and control various areas of social reality.

The great majority of studies on metrics in academia link the use of metrics and their repercussions with the implementation of processes oriented toward a neoliberal and knowledge-based economy. Nonetheless, as I argue in Chapter 3 where I discuss modernization processes in Russia, the metricization of academia has been rooted not only in demands for accountability but also in the modernization of the science and higher education sectors for the benefit of developing the whole nation.

Metricization consists in the implementation and use of metrics (more specifically: measures, metrics, and indicators) for monitoring, reporting, evaluating, and managing. Through the implementation of metrics in academia, the evaluative power of global actors (e.g., international organizations like the World Bank, producers of bibliometric indicators etc.), national actors (e.g., governments, national funding agencies), and local actors (e.g., university management, deans) is introduced and reproduced. Metrics represent an assertion of (evaluative) power to produce knowledge about certain fields and to shape the way in which these fields are understood and governed. Thus the main motivation for the introduction of metrics is to reduce the use of the most valuable resources such as time, money, and expertise in decision-making. Metricization plays an important role in economization, particularly in the introduction of market-oriented reforms of the higher education and science sectors. In this way, individuals and organizations are constituted as economic actors and entities whose economic character (*homo economicus*) is reproduced by practices of measuring, reporting, controlling, and assessing. The dominance of economization over metricization is established through power relations: measures are hallmarks of economic (although not exclusively) power.

Metrics are not only employed in order to make decisions simpler. Metrics are created and used to shift the burden of responsibility for the decision that has been taken from those who use metrics to the metrics themselves. By using metrics, decision makers increase their own legal or moral authority. This is due to modern society's trust in numbers as representations of measured worlds (Porter, 1995; Rose, 1991). Viewed as objectivized representations of social reality, metrics are treated as neutral, as are the media, for example. In discussions about metrics, however, one too often forgets that metrics are socially constructed and as such cannot be neutral. Their form is the result of the play of various powers; thus they represent not so much an objectivized social reality but rather a way of seeing and understanding it that has been constructed. Thus, taking up Kula's perspective once more (1986), one can argue that metrics arise as a result of social conflict between those who want or need to measure and those whose work – or who themselves – are measured.

Although they are ontologically neutral, metrics are politically determined because the choice of the underlying value of logic, which metrics serve, is a political action. Relations between metrics and politics, as well as policy, are mutually constitutive: political and policy decisions depend upon metrics yet at the same time, acts of quantification and metricization are politicized. Expressed in numbers, metrics achieve the privileged status of objectivized tools, through which fact they start to be perceived as de-politized and autonomous. Designers and users of metrics try to conceal the social character of their construction and the social conflicts that determine the final shape of metrics. In this way, for those whose are measured, metrics can become a goal instead of a measure. Those who design metrics assume that they can provide incentives to improve performance; what they most often produce, however, is the aspiration to achieve better metrics expressed by higher numbers: Ex post metrics are translated by those who are measured into ex ante targets. Occasionally, metrics as measures mapped into numbers are very weakly related to actual performance. Metrics are designed to measure and improve performance, but often the relation between these two concepts is not ontological but political (i.e., metrics designers can impose a given metrics in order to measure and change a given social phenomenon or practice).

Gläser and Laudel (2007) have analyzed the tendency to make metrics autonomous in the field of research evaluation. In their constructivist approach that is based on Bruno Latour's writings, they show that "modalities," that is statements about statements that modify the validity and reliability of statements about scientific findings, are dropped when bibliometric indicators are used in evaluation. Modalities might be understood as signaled limitations of any metric, and as calls for contextualization each time a given metric is used. For instance, professional bibliometricians and members of the scientometrics community draw attention to, among others, the following modality, that the database of publications used in evaluation changes over time, and this change must be taken into account in evaluations. Moreover, publications must be assigned unambiguously to authors and their organizations. The black box that is created when modalities are dropped can be observed, for instance, in the definition of research excellence presented in the European Commission (Directorate-General for Research and Innovation, 2014) report, in which information on data sources is provided (e.g., for counting publications the Scopus database is used), but only the experts in bibliometrics and scientometrics actually know what the consequences are of choosing a particular source for the measurement of results.

Looking into measurement practices across a variety of disciplines, one sees that in evaluation exercises, designers of evaluation systems and evaluators use databases as if they reflect scientific reality: The fact that some fields of science are not indexed and reported in a given database means that they do not produce knowledge, and not that the database's scope is insufficient. In addition, bibliometric indicators that lack explicitly stated modalities are de-contextualized and easy to use. They are easy to use both for evaluators and the public and, moreover, such indicators attract more publicity than their interpretations. Discussing the role of numbers in scientometrics and outlining the pitfalls of metrics-based models of research evaluation, Marewski and Bornmann (2018), highlight five problems resulting from the ways in which numbers are used: (1) the skewedness of bibliometric data, (2) its variability (3) the time and file dependencies of bibliometric data, (4) the domination of English bibliometric data, (4) and the coverage and incompleteness of databases. Where they become the dropped modalities in the sense presented by Gläser and Lauder, these become grave problems.

To take an example, one could compare the number of citations a microbiologist versus a philosopher achieve. This would be a straightforward comparison between just two figures, and it would thus appear as a legitimate exercise. And yet in this case, one would need to take into account the fact that publication patterns in the biological sciences and humanities differ, resulting in a higher number of multi-authored publications in microbiology and, as a consequence, in a higher number of citations. One has also to consider the fact that the majority of international bibliographic databases are lacking in their coverage of publications from the humanities and social sciences (Bonaccorsi, 2018a; Kulczycki et al., 2018; Sivertsen, 2014). Thus the use of "objectivized metrics" in the black box of research evaluation, when it occurs without a highlighting of modalities, covers up different social conflicts and intensifies, often in an unnoticed way, different power games.

The rise of metricization and the use of metrics for managing and evaluating science has its own history that is often presented from the Global North's perspective, which highlights the role of accountability and of New Public Management. In this narrative, the founding of scientometrics by, among others, Price (1963) and Garfield (1955, 1964), constituted an effort not only to understand and describe science but what also became a policy tool for auditing, monitoring, and evaluating science. The key measure in such scientometrics is an individual scientist whose work can be aggregated into the work of an institution, country, or region. By so doing, a government or evaluation agency can assess the "excellence" or "productivity" of a given scientist through various indicators like the number of grants or publications, or the amount of external funding generated. In such a history, one could draw a straight line from the foundation of scientometrics in the 1950 and 1960s, through the emergence of New Public Management and the knowledge-based economy in the 1970 and 1980s, to the rise of the first national research evaluation systems in the 1980 and 1990s and the overwhelming presence of scientometric indicators in the twenty-first century. Such a neat account is enticing, which in fact explains the prevalence today of this kind of narrative. However, I would like to propose a different history of research evaluation systems. In Chapter 3, which describes the role of metrics in higher education and science in Imperial and Soviet Russia, I show how certain paths to modernization, and therefore trajectories of research evaluation systems, have been overlooked.

3

Untold Histories of Research Evaluation

A few years before he attended the 1951 Congress of Polish Science, held during Stalinism's apogee, Frédéric Joliot-Curie visited Lavrenty Beria, the Minister of Internal Affairs of the Soviet Union, and asked him to save the imprisoned Nikolay Vladimirovich Timofeeff-Ressovsky (Vitanov, 2016). Timofeeff-Ressovsky was a famous Soviet geneticist arrested in October 1945 and accused of collaborating with Germany during World War II. Thanks to Joliot-Curie's intervention, Timofeeff-Ressovsky was rescued, although by that point he had almost died of starvation and was nearly blind (Ings, 2016). Although the Soviet geneticist was saved, he was not released but rather transferred to Object 0211: one of the popularly called *sharashkas*, that is a secret research and development laboratory within the GULAG. The Soviet system recognized the importance of Timofeeff-Ressovsky's scientific work; thus he was forced to work in Laboratory B that was participating in the Soviet atomic bomb project.

When Americans detonated two nuclear weapons over Japanese cities in August 1945, Stalin was forced, to his great dismay, to recognize that the Soviet Union was lagging behind the US Army in military terms. It was for this reason that the whole system of research institutions, secret GULAG laboratories, and the highest authorities of the Communist Party got involved in planning and managing the Soviet atomic bomb project. The Soviet Union was perceived by the Americans as incapable of building nuclear weapons so soon after having invested so much in the war effort. Yet they quickly began to realize how effectively the Soviets were in their gradual infiltration of the Manhattan Project.

The Americans were convinced that they were the richest, most advanced, and most technologically mature country, many years ahead of the Soviet Union in terms of military development. Therefore, when Sputnik, the first artificial satellite, was launched on October 4, 1957, the Americans' self-importance collapsed. Not only had the Soviet Union won the first round of the space race, more importantly, it had showed the Americans that by developing its space program and a new generation of rockets, the Soviet Army could carry thermonuclear weapons over American territory. Sputnik, which was equipped with radio transmitters, could be seen in the morning before sunrise and in the evening just after sunset with simple binoculars, and broadcast signals could be picked up even by amateur radio enthusiasts (Johnson, 2017). This was then a visible and terrifying proof for everyone of Soviet domination of space.

The race into space pit capitalism and communism against each other. The Americans understood that they had to modify the organization of their research landscape. One year after the launch of Sputnik, the National Aeronautics and Space Administration (NASA) was established, and two years after Sputnik, the Advanced Research Project Agency (ARPA), which was responsible for the development of emerging technologies for the army, was launched. In 1958, which was a milestone in the history of American science policy (Killian, 1977; Neal et al., 2008), President Dwight D. Eisenhower signed the National Defense Education Act which aimed to train defense-oriented personnel and encourage young people to become engineers. Scientific development that contributed to the space race began to be more centrally planned and managed, which in fact, to a certain extent, made it resemble elements of the Soviet approach.

Most of the studies on Sputnik focus on it as part of the story of the space race, detailing its impact on the US scientific policy. Little attention is paid to the system through which science was organized, which enabled the Soviet Union, a country exhausted by the war effort, to launch the first artificial Earth satellite. The history of central planning and the accounting of research initiatives often goes untold, because in the dominant perspective produced by the most scientifically developed countries today, all attention is focused on the transformation of American science policy under the influence of Soviet domination in space, and the collapse of the Soviet system (see, for example, Killian, 1977; Plane, 1999). The system collapsed decades after the launch of Sputnik which fact is explained as, among other things, a product of the absence of pluralistic market forces. Such a focus, however, leaves unanswered the question of how the Soviet system of science was able to achieve such impressive scientific results that also had an impact on its economic growth. One of the key factors was a national ex ante research evaluation system that significantly contributed to Soviet research achievements and sustained its science as a key modernization tool of the Soviet system.

* * *

Through an examination of modernization processes, in this chapter I investigate why certain aspects of the history of research evaluation systems have been overlooked. Following from this, I argue that it is crucial that we research these areas if we want to understand contemporary research evaluation games' forms and effects. I start by showing that modernity has multiple faces and that some of them, thanks to diverse geopolitical narratives, are neglected. Modernization processes are characterized by certain universal logics like metricization and the primacy of economic value as a criterion for measuring success. These features, however, become distorted during the implementation of modernization projects in new contexts; with this diversity constituting the varied trajectories of modernization. In elaborating the argument, I describe two such trajectories, the so-called capitalist and socialist ways. The former is a dominant grand narrative that focuses on the rationalization and Westernization of the social world. The specific character of the latter is a result of revolutionary and, most significantly, accelerated transformation of an existing class structure, bureaucracy and administration through the use of central planning and the deployment of tools based on quantification, metricization, and science. One can examine this socialist way – overlooked in the history of scientometrics and research evaluation – in Russia where, from the establishment of the first Soviet state in 1917, it began to mature.

It is through an investigation of this neglected aspect of research evaluation history that one can understand why, within similar evaluation regimes, the evaluation game can assume different forms. My exposition of the specificities of the socialist way, which contributed to the emergence of the first ex ante research evaluation systems a century ago, is focused on two main themes: (1) central planning in research and in the reorganization of the science landscape in the Soviet Union and socialist countries and (2) the rise of scientometrics in the Soviet Union as a program for the management and understanding of science. My aim is not to provide a brief history of each of these themes but rather to present their key characteristics. This then allows me to elaborate two models ("neoliberal" and "socialist") of research evaluation systems in the final part of this chapter.

Almost since the birth of modern science in the seventeenth century, research evaluation has been an integral part of research itself. The scientific revolution was based on two main pillars: the verification of scientific knowledge through the observation of experiments and the communication of results of experiments to enable an independent and critical confirmation of discoveries. The confirmation of results in fact constituted an evaluation of research that served to verify whether research followed scientific methods. The increase in the number of scholars and experiments required a firm basis for the discussion and evaluation of research results. Thus the first scientific journals were established at the end of the seventeenth century (Fyfe et al., 2019; Moxham & Fyfe, 2018). Today, the birth of scientific journals is linked with the establishment of peer review practices but, as Baldwin (2018) shows, it is likely that these practices emerged later. The concept of peer review does however date back to 1731 when the Royal Society of Edinburgh established a practice of distributing correspondence to those members who were most versed in the matters raised in the correspondences (Chapelle, 2014). Moreover, despite the professionalization and institutionalization of science from the nineteenth century onward, peer review was not a standard practice in publishing scientific journals until the mid-twentieth century (Baldwin, 2018). The massive growth in the number of researchers, institutions, and publications generated the need for bureaucratic management of this part of the social world. The scale of the demand for experts and reviews made it necessary for peer review to be supported by additional tools. In this way, qualitative evaluation (peer review) began to be supported by quantitative evaluation (metrics), and over time, has in practice been displaced by it in various areas like university rankings.

Both universities and research institutions, conceived of as tools of modernization, on the one hand, use metrics to control, monitor, and develop various social practices and, on the other, are measured and controlled through various metrics and indicators. Economization and the metricization of academia are not, however, unique to the Global North. This is because neoliberalism is not a prerequisite for establishing metrics as tools for the management and assessment of science. As I show, modernity has multiple faces of which one, that is the Eastern, is often disregarded. Where one's key aim is to understand the effects of currents systems of research evaluation, one must look at the context in which they arise. Doing so, one must pay attention to this overlooked aspect of history, which one can define as consisting of two key dimensions, that is, central planning and the rise of scientometrics in the Soviet Union. Both dimensions are described in the literature on the history of science and research in the Soviet Union, but there are mostly omitted in the history of scientometrics and research evaluation. Here I present their key characteristics, which allows me to delineate the ideal types – in Weber's meaning of the concept – of research evaluation systems established in the West and in the (Global) East. Next I take a closer look at these two themes in order to compare approaches to research evaluation systems stemming from metricization and economization in various geopolitical contexts.

3.1 The Multiple Faces of Modernity

This chapter aims neither to explain the modernization of societies through their bureaucratization and rationalization, and the implementation of the neoliberal model of globalization, nor does it set out to show that only in a "Western society" could research evaluation systems emerge. Rather, this chapter strives to show that there is more than one modernity because different countries are situated in different positions within the global network of power relations. That is why modern states have established at least two ways of finding and sustaining their place in this network. I refer to them as capitalist and socialist ways or trajectories. The difference between these trajectories is a product of the acceptance by various actors (states and their political institutions) of different logics and values which underpin the design of diverse technologies of power.

It is critical that we premise that there were two ways of establishing a position in the global network of power relations. Only in this way we can paint a comprehensive picture of the processes that laid the grounds for research evaluation systems not merely in countries like Australia or the UK but also in Poland or Russia. Approaches based on classic modernization theories are not up to the task of describing the context in which research evaluation systems arose because they disregard some regions of world and devalue modernization pathways different from those of the Western countries. This has produced a gap in knowledge about socialist and post-socialist research evaluation systems. This is due not so much to the research choices of individual scholars but rather to deeper, systemic causes that can be identified in three intertwined phenomena: (1) oversight of alternative ways of understanding modernity; (2) the invisibilization of the Second World after the fall of the Iron Curtain through the transformation of the grand narrative from that of an East–West division of the world to one describing the Global North and Global South; (3) a disregard of peripheral countries coupled with a failure to recognize that knowledge production in such countries functioned as a tool of modernization tool that accelerated social and economic development.

Modernization theories, notably those based on the thought of Max Weber (1978) and Parsons (1987), outline a universal course for modernization processes. It is only in the movement of critical interpretation that we see that such theories are in fact limited to what was and is happening in the West. In other words, these theories serve to describe how much a country has become a *Western* country. Nevertheless, it is neither possible nor desirable to completely avoid concepts used in varied discourses on the modernization of societies and countries. My perspective owes much to Immanuel Wallerstein's world-systems approach (2004) and Martin Müller's (2018) recovering of the Global East. These allow us to conceptualize global history in a manner that is not restricted to the idea of capitalist modernization, but rather perceives it as a history of power relations across core, peripheral, and semi-peripheral countries. Following the approach developed by Müller in a polemic directed at Wallerstein's perspective, one can reconceive of these power relations as being defined across the Global North, Global South, and Global East.

Let us now turn to the three phenomena mentioned above in order to clarify why the socialist way has been overlooked and why including it in a history of research evaluation systems will enhance our understanding of the use of metrics in science.

I examine the socialist trajectory for two reasons. First, because it is itself an important part of the history of research evaluation. Second, because it can shed new light on our understanding of the Western (capitalist) path of modernization.

(1) The Omission of Alternative Ways of Understanding Modernity

If a history of research evaluation systems takes as its starting point the systems that currently operate, there is a danger that it will base itself on an epistemic error. As research evaluation scholars, we trace the origins of contemporary systems in core Western countries like Australia, Denmark, Norway, and the United Kingdom, as well as in other semi-peripheral, but still "Western" countries like the Czech Republic, Poland, or Slovakia. Then we analyze the socioeconomic contexts within which such systems were established (during the 1980s and 1990s). We duly draw the conclusion that all of these systems are the product of global economic struggle, and that they have served as tools for implementing neoliberal solutions to the economic and social problems faced by Western states during the last decades of the twentieth century. It is no wonder then that capitalist modernization theories force us to reduce our understanding of research evaluation systems to systems based on neoliberal grounds.

I believe that we should take a different approach to writing the history of research evaluation systems and reconstructing the context and conditions that enabled them to emerge. In order to do so, I propose that we start not from current systems, but rather from the beginnings of when metrics were used in planning *for* and *of* science as part of emerging modernization processes. Investigating how academic work and research were organized through plans and numbers and how academia started to be managed and deployed for the modernization of societies and emerging national states enables us to move beyond an account of the origins of capitalist path modernization and the use of neoliberal tools for the Westernization of science. Instead, it allows us to also trace a distinct socialist path to modernization and to investigate how research evaluation was used as one of its tools.

Although to this day most mainstream research on higher education and the science sector accepts the universality of a single Western European (in other words, Global Northern) model of modernity and development through capitalism, this approach is slowly being challenged. For instance, Eisenstadt (2000) shows that "multiple modernities" exist, and one of the key implications of this idea is that "modernity" is not equal to "Westernization." To explain political, societal, and cultural processes with reference to Westernization constitutes a type of symbolic violence and valorization that assumes that other non-Western models of modernization are *ex definicione* non-modern or archaic. We can see this approach at work in the foundational texts of Loren R. Graham (1990, 1993), the American historian of Soviet and Russian science. In the 1920s, the government of Soviet Russia and scientists were elaborating a highly sophisticated model of science development through which to accelerate the country's modernization. This model of a scientific organization of science and research was based on a tradition of using metrics initiated, as I detail below, nearly 200 years earlier. Thus, by 1920s, Soviet Russia inherited a long history of modernization through science. At yet, central planning, even in the works of Graham, who published extensively on this topic, is routinely assessed as a function of its level of Westernization.

The modernization of Russia initiated by Peter the Great in the seventeenth and eighteenth centuries developed along its own paths, independent from Western ones, and impacted other countries like Poland and the Ukraine. It laid the foundations for the "scientific organization of scientific labor" by which the Soviets aimed to increase researcher productivity either by improving research methods or material conditions in researchers' workplaces. The process begun by Peter the Great laid the basis for an alternative socialist path of modernization with its own Russian method and brandished the flag of the French Revolution's ideal of the fraternity of all mankind. The epitomes of this path were the Russian Revolution of 1917 and the founding of the Soviet Union in 1922. Peter the Great's reforms inspire reflection on the universal and particular dimensions of modernization, because in them one can observe both the common roots or features of modernization, and their subsequent transformation when they were brought to Russia. These reforms were ultimately very "Western" because it was his journey through the West that inspired Peter the Great to introduce them, using "Western experts" of the time. However, the logic of these reforms, imported as they were from the West, was deformed in the particular context in which they were implemented.

One of the pillars of the Russian modernization project was the establishment in 1725 of the Imperial Academy of Sciences, which transformed over time into the Soviet, and later Russian Academy of Sciences. As Graham (1967) shows, thanks to the 1747 charter, academics were required to provide plans of their work in advance of their research. This was an annual obligation which was followed up by the submission of reports at the end of the year. Further, the practice of measuring researcher productivity in terms of the number of publications produced is not, as is commonly thought, a heritage of New Public Management (NPM) or the neoliberal university. It can instead be traced back to the 1830s, when Russian university professors were first obliged to publish a paper every year (Sokolov, 2016).

In *Evaluation Society*, Dahler-Larsen (2012) argues that evaluation is a product of modern society and as such reflects the norms and values of the environment that shapes it. If as scholars investigating research evaluation, we agree, then we must also recognize that different modernization paths and traditions of understanding modernity, will play a highly determinant role over the form and content of evaluation in science. Thus while in contemporary Australia, Poland, Russia, and the UK, we might observe the use of the same policy instruments, our understandings of how they work, for what reason they were implemented, when they were designed and by whom, must vary. Research evaluation is a product of modern society. However, there have been multiple modernities that have shaped different context for evaluation in countries influenced both by capitalist and socialist modernization.

(2) The Invisibilization of the Second World and of Socialist and Post-Socialist Countries in the History of Modernization

After the Cold War, thinking about global differences and geopolitics started to be dominated by the distinction between the Global North and Global South. The categories of the North as representing wealthy and developed countries and the South as representing poor and underdeveloped countries have replaced the division between the so-called First, Second, and Third Worlds. Northern countries are perceived not only as wealthier but also perceived as producers of guarantors of the knowledges that are transferred to the countries of the Global South.

Moreover, since the fall of the Communist regimes, the grand narrative of the Western versus Eastern division of the world as the division between capitalism and communism is no longer used. Reuveny and Thompson (2007) show that the First World became the North and the South replaced the Third World. This has led, however, to the Second World, linked to communist and socialist societies and countries, becoming invisible (Grzechnik, 2019). Müller (2018) shows that ex-Soviet, former USSR, and old Eastern bloc countries continue to be considered "former communist bloc," and yet, have not managed to secure their place in the North–South distinction which in fact erases them. Thus, Müller (2018) proposes that we use the concept of the Global East to put "the East back on the map of knowledge production" (p. 3). Following this suggestion, Kuzhabekova (2019), writing about Kazakhstan, argues that keeping Euroasia uncategorized is beneficial for neo-colonial regimes. I agree with Müller that the East should be conceived of as an ontological and epistemological category rather than a geographical one. In this way, the East provides identity to the Other (other to the societies and countries of the North and South), and there is no need to introduce the category of a new West. The Global East has always existed, but we need the concept and term in order to make it visible in our discussions.

While in the past, post-socialist states belonged to peripheral and semi-peripheral regions, today, for instance, Central and Eastern European countries in the European Union are economically developed and therefore closer to the core than to the semi-peripheries (Zarycki, 2014). However, they are perceived neither as countries of the Global North nor of the Global South. This invisibilization of the Global East makes it impossible for us to formulate knowledge about the condition of countries and economies which for many years followed the alternative path to modernization. Further, this disappearance is not benign in terms of its consequences for research (especially in the framework of social sciences) on post-socialist countries.

Thinking through the categories of the Global North and South reveals a blind spot in knowledge about research evaluation systems in the Global East. And yet an investigation of the Global East can be very illuminating as socialist modernization led to the emergence of research evaluation systems there much earlier than they appeared in the Global North. Understanding this context is crucial for identifying how the evaluation game appears and functions in the Global East, and why the same global policy instruments provoke varied resistances and adaptations at the local and institutional levels.

By restoring the Global East to its place in the history of modernization, investigation into research evaluation systems can understand why the neoliberal policies of measuring science imposed on post-socialist systems have met with resistance and hostility. Thus on the one hand, neoliberal policies appear similar to historically earlier ways of organizing science systems in these countries, and, on the other hand, they conflict with historically established patterns of development in the science sector.

A key consequence of rendering the modernization history of the Global East invisible is that we use conceptual tools produced mostly in the capitalist Global North to understand past and current science and higher education systems in the Global East. In this way, we miss the entire context of alternative paths to modernization with their own logics distinct from those of capitalism. Thus although there are many similarities among the policy tools designed for improving research performance in the Global North, South, and East, they do not reflect an identical logic because the contexts in which they arose were shaped by multiple modernities.

(3) *The Failure to Recognize that Knowledge Production Was a Tool of Modernization in Peripheral Countries Designed to Accelerate Social and Economic Development*

From the perspective of world-systems theory, the basic unit of social analysis is the world-system rather than the nation state (Wallerstein, 2004). A world-system is a socioeconomic system defined by the existence of a division of labor. The modern world-system has a multi-(national) state structure consisting of three zones related to the prevalence of profitable industries or activities: core, semi-periphery, and periphery. Countries may be defined as core or semi-periphery on the basis of their position in the process of the redistribution of resources. Typically, resources are redistributed from peripheral countries (most often the key resources are raw materials) to industrialized core countries. According to Wallerstein, the beginnings of the modern world-system can be traced back to the "long" sixteenth century (1450–1640) when Latin America and Eastern Europe underwent processes of peripheralization. In the current moment, it is both fruitful and important that we investigate this history through the lens of the core–periphery division. The historical experiences of areas and countries pushed into peripheral positions makes it possible to understand the constraints currently faced by these states in their attempts to modernize semi-peripheral zones. Countries – as Babones and Babcicky (2011) argue in relation to Russia and East-Central Europe – are strongly bound to particular positions (in the three zone structure) by long-lasting social, cultural, and geographical ties.

It was the constraints of being a peripheral country – that pushed Russia into rolling out an alternative modernization trajectory that had a great influence on the organization of science and higher education both there and in the Soviet Union and in other countries influenced (culturally, politically, and militarily) by it. Since the foundation of the Imperial Academy of Sciences in 1725, science in Russia was treated as a branch of government and subjected to imperial command (Graham, 1967). The role of scientists and academics was to cultivate and develop native Russian science and, through numerous expeditions, explore remote parts of the empire. It is also worth mentioning that the effectiveness of the Academy established by Peter the Great was in evidence only in those areas of science that were locally significant. The charter prescribed the development of those sciences that were useful to the state. Therefore, the contribution of the Russian Academy to world science in the eighteenth century was very minor (Dmitriev, 2016). The academy and later the universities were treated as the key tools of modernization which were centrally governed and which had to document and report on their activities. Such bureaucratic tools had to be designed and implemented in Russia because in the eighteenth century, it had not reached as significant a level of bureaucratization as was found in Western countries. Thus, the bureaucratization of management processes initiated with Peter the Great's reforms propelled Russia onto the path of modernization.

Peter the Great's inspiration came from the West, and his reforms clashed with Russian heritage and tradition. As a result, both the tools of modernization in Russia and its entire modernization project took on a different character than it did in the West. The modernization of Russia through figures and tables, documentation and reporting does not reflect the same social or cultural impulses that were the drivers of modernization in Western European countries. Modernization through quantification has universal origins that might be observed, for instance, in both France and Russia. And while these countries share common experiences, Russia became modern through a socialist and not capitalist path to achieving the ideal of the fraternity of all mankind. This socialist way, which, like its capitalist counterpart, stemmed from the universalist ideals of objectivity and measurement, took form through the establishment of the first Soviet state in 1917.

Within neoliberal approaches to science, it is assumed that metrics or key performance indicators serve to express what a given institution wants to achieve, and that they result from practices rooted in NPM. However, as noted above, Russia was the first country to devise and implement formal research performance indicators as state policy tools (Sokolov, 2021) – a full two centuries before NPM emerged. From the 1830s, university professors were required to produce journal articles on a yearly basis and their salaries depended on the number of papers they published. These regulations were a product of the earlier introduction of nationwide rules for documenting and reporting on official activity in the country. In the 1720s, as Emysheva (2008) shows, Peter the Great started work on the ordering of the Russian Empire's document management system. The question of the technology of documentation was given such an important place in the administrative reform's key legislative act because document management was identified with management activity itself. As some researchers show, for example, Anisimov (1997) in his book on state transformation and Peter the Great's autocracy, the bureaucratic system was designed to replace the role of the hereditary nobility in the state by a centralized system of ranks. This system was based on documents and reports. The 1720 General Regulation became a prototype for current strategies for the management, registration, and archiving of documents, as well as for the design of forms. The introduction of the new system was supposed to support the development of the army, economy, industry, and culture. Peter the Great stressed the idea that thanks to the changes he introduced, the previous way of making decisions on an individual basis had been pushed aside in favor of collegial decision-making. The 1720 General Regulation regulated the entire process of document management: from the receipt or creation of a document to its delivery to its recipient or to the archives. One of the key points of the regulation was the definition of standards for the documentation of management activities. It mandated that there was an obligation to document (and thus to report on) activities related to the management of any public activity. The storage of documents was also developed within it and the regulation outlined a plan for the establishment of the entire system of state archives. This system became the prototype for other systems regulating the activities of contemporary offices in terms of organization and methods.

When Alexander I established the Ministry of National Education in the Russian Empire, he inaugurated the era of reporting and documenting work not only in education but also in universities, the academy of sciences and learned societies

(Galiullina & Ilina, 2012). At the beginning of the nineteenth century, Russian universities started to publish their own scientific journals. As Galiullina and Ilina (2012) show, the Imperial Kazan University required its employees to publish their research results in the Kazan University journal. This was a consequence of the modernization process in which Russia recognized universities as centers that could produce ethnographic scientific information and descriptions of the whole expanse of Russia's territories. In this way, universities became key players, alongside the Russian Academy of Sciences, in the system of producing scientific knowledge.

In the years 1832–1833, as Vishlenkova and Ilina (2013) show, the Russian ministry sent all institutions the forms on which employees had to report their activities. All information provided by university professors had to be confirmed by other persons. Vishlenkova and Ilina (2013) draw attention to the fact that the characteristic feature of the Russian academic management system was that officials required university staff to use a reporting system analogous used in to state offices, to follow typical official rules, and to use official language. As a result of a more aggressive implementation of this document management system, university employees were forced to become officials. At the same time, the burden of reporting every activity and compiling statistics was very heavy for academic staff members. As Vishlenkova (2018) mentions, famous mathematician N.I. Lobachevsky must have taken part in preparing documentation for 300 meetings in 1838 alone.

Galiullina and Ilina (2012) describe another interesting facet of the history of Russian universities. In 1833, the ministry created its own scientific journal, in which each employee of the university "could" publish at least one scientific article. This decision by the ministry was perceived by the academic community as a sign of the upcoming selection of academic staff. Moreover, each scientific institution was obliged to provide the ministry with a list of potential authors who would send their articles to the journal in the future. These radical changes in the system of science management caused panic in the scientific community. The universities created lists of potential authors who would undergo rigorous evaluation by a minister who was famous for his relentlessness. Because the academic community considered this regulation as a threat and a tool for staff selection, the universities, employing extra caution included in the list of potential authors the entire staff of each research unit. Interesting in the light of today's discussions on the societal impact of research is the fact that the quality of a potential scientific publication was assessed on the basis of its usefulness, and therefore, authors described this usefulness at the start of their articles. To this day in Russia, in all student, postgraduate, and research work written in Russian, one has to elaborate on the study's relevance in the text's introduction.

As a result of these changes in reporting on and management of universities, for the first time, scientific activity became the evaluation criterion for members of the academic community. As Vishlenkova (2018) argues, complying with evaluation regulations rendered university professors fully dependent on the system. Every activity that was outside of the template could not be properly reported and was thus classified as marginal. Thus, the criteria determined the areas that actually counted for the Russian ministry, and everything that could not be reported was perceived as a redundant activity. It is noteworthy that the ministry's officials produced statistics on the publication activity of university employees and published them in the form of tables every year, which made it possible to demonstrate the local achievements and character of each university. Standardized tables and forms allowed the ministry to combine diverse types of information from across institutions into a single numerical figure that reflected the quality of those institutions.

As Sokolov (2016) describes, the practice of compelling university professors to produce a publication every year, first enforced in the 1830s, continued through most of late imperial and Soviet history. The planning of the science sector and the further spread in the late 1940s and 1950s of Stalin's model of science and higher education organization made science management and evaluation, unheard of in Western science in those days, permanent features in Russia (Graham, 1967, 1993; Sokolov, 2021). There, where central planning *for* and *of* science was both discussed and implemented, from the very beginning of the twentieth century, there were no questions as to whether science could be measured and evaluated. Instead, the question was rather *how* it should be measured in order to provide the best results for the state and economy.

As shown above, in early modern Russia, knowledge production was viewed as a tool for modernization, and its aim was to speed up social and economic development. It therefore developed mechanisms for controlling and monitoring this area, and for measuring and evaluating knowledge production on a national level, much earlier than its counterparts in the West (Global North). A lack of knowledge about socialist modernization, which in this book refers mostly to Russia and the Soviet Union, leads to inadequate knowledge about socialist and post-socialist research evaluation systems whose origins, as I have shown, are neither Western, nor neoliberal.

3.2 The Central Planning of Science

From the mid-nineteenth century then, numbers became tools for the planning of science in Russia. *Ученые записки* (*Academic Notes*), founded in 1832, was the most important scientific journal of the time (Galiullina & Ilina, 2012). From the very beginning, the editorial board planned the exact number of author sheets (one author sheet is equal to 40,000 characters with spaces), which would constitute the issue. For the first issue, more than double the anticipated number of author sheets were published as demand far exceeded the expectations captured in the plan. Almost 100 years later, outdoing the plans and performing the 200% standard would become an important narrative in planning not only of science but also of the economy and other areas.

In the current literature on the use of scientometric indicators for increasing research productivity, the common view is that such indicators began to be used in science policy during the 1970s, for example, by US universities in their decisions on promotion and tenure and by the US National Science Foundation (Gläser & Laudel, 2007). At that time, scientometrics and bibliometrics indicators were starting to be used outside of academia and the bibliometricians' community to monitor and assess the science sector. It should be noted, however, that both in Europe and the United States, researchers themselves had begun to use indicators to analyze the development of their disciplines already at the turn of the nineteenth and twentieth centuries (Godin, 2009; Haitun, 1980).

Before the idea of central planning of research spread across Russia, the scientific organization of scientific labor (научная организация научного труда [NONT]) had already been initiated by Alexei Gastev, a disciple of Frederick Winslow Taylor, the father of scientific management (Graham, 1967; Lewis, 1979). Founders of NONT believed that the productivity of research could be increased by improving scientists' work methods and the material conditions of their labor. Graham (1967) describes an idea of N.P. Suvorov, one of the NONT theoreticians, that he presented in the 1928 edition of *Scientist* (*Научный Работник*) journal. It consisted in an attempt to compute the effectiveness of research by means of algebraic formulae. According to Suvorov, using this method would greatly facilitate science planning because through it, the most demanding tasks could be assigned to the most talented researchers. Suvorov showed how the formula could work by using the example of the eighteenth-century scientist Lomonosov. NONT theoreticians assumed that researchers had to be evaluated in certain cycles (which varied for different researchers), and which represented the time needed for a scientist to produce a major scientific discovery. Suvorov suggested the following formula: $T = At + 3B + C/X + D/2X$, where T = "effective time," t = the number of years of the report cycle, and X = the average number of "regular" works a year. A, B, C, and D designate numbers for four types of work: (A) of great scientific significance, (B) outstanding, (C) regular, and (D) not demanding independent analysis. According to this formula, in the years 1751–1756, Lomonosov did 33 "effective" years of work in this six-year period. Although this formula was assessed by Graham (1967) as "utopian and naïve" (p. 51), almost a century after its publication and 60 years after that of Graham's book, it might look very familiar to university managers and researchers. All one has to do is place articles in *Science, Nature* or *Cell* beside A, articles in JIF journals beside B, scholarly book publications beside

C, and reports beside D. In this way, one can get an extended version of an author impact factor (Pan & Fortunato, 2014) or the sum of weighted points used in various performance-based research funding systems such as in Denmark, Norway, Finland, or Poland (Pölönen, 2018; Sivertsen, 2018b).

The Russian Bolsheviks regarded as axiomatic both the Marxist critique of capitalism and the idea that the market economy was inherently inefficient and therefore unfit for coordinating large-scale industrial production (Ellman, 2014). The superiority of planning, through which production could be coordinated ex ante by society as a whole, became a widespread idea. In relation to this, the origins of central planning in science as well as research, development, and innovation can be tracked back to the plan for the electrification of Russia in 1920, which was intended to increase economic growth. However, the central planning of research in its most well-known form began to be developed in the late 1940s. According to key persons involved in Soviet science policy, capitalist societies were not able to implement central planning on a national scale. The obstacles were supposed to be: the varied sources of financial support for research and, particularly emphasized, the use of research results by diverse social and economic groups with conflicting goals. In Soviet society, on the contrary, the whole society constituted a single group, a single class, which shared one common goal: the development of Soviet culture and economy (Hare, 1991).

Today, in neoliberal approaches to evaluation, it is assumed that research evaluation at the national or global levels should be focused on institutions (e.g., by university rankings), which have autonomy to define whatever research topics they deem appropriate. Within this perspective, the results of research should be assessed according to their excellence. This institutional ex post approach is not, however, the only way of evaluating research at the national level. Decades before the establishment of the first national research evaluation systems in the Global North and the first university ranking, the Soviet Union set up a national ex ante evaluation system focused on assessing research themes or programs within a particular institution before they were conducted.

The Soviet Union and other Eastern Bloc countries implemented central planning in science with a number of goals: In order to achieve high production growth rates, to make more efficient use of investments in the national economy, to enable the development of all branches of the national economy by raising the scientific and technical level of research, and to ensure continuous improvement of people's material and cultural values (Kapitza, 1966). To achieve these ends, in cooperation with ministries and industry federations, the State Planning Committee (Gosplan) in Soviet Russia implemented the planning of *topics* (or tasks), that is orders for the conduct of relevant studies of interest to the national economy. In 1948, S.I. Vavilov, the president of the Academy of Sciences of the USSR, highlighted that "the plan of scientific development in a socialist state must, of course link up with the state economic plan; but (…) science has its own peculiar logic of development, a logic which it is essential to take into account" (Vavilov, 1948). However, the logic of science had to be subordinated to the Soviet economy and the idea of the New Soviet man (i.e., a person who is selfless, learned, healthy, and in favor of spreading the socialist Revolution) produced by the socialist state.

Before the 1950s, the planning of science as well as research, development, and innovation were treated as part of the economic plan. After the launch of Sputnik, when the importance of the race between the United States and the Soviet Union for scientific primacy intensified, science planning was assigned to a separate sphere of planning. In the 1970s, new principles of "program-goal planning" were introduced to provide broad and long-range economic–social–technological goals to achieve through planned research (Nolting, 1978). Moreover, measures were introduced for different branches in order to realize projects under a single inclusive program.

Bukharin – one of the most influential Soviet theoreticians – contended that the planning of science in capitalist countries took place not at the national level but in each laboratory in which planning was implemented (Graham, 1964). For Bukharin, the planning of science was not only a way to improve the economy but also to increase the productivity of science itself. One of the pillars of planning was careful calculation and book-keeping, by means of which the whole system could be organized. Book-keeping offices or statistical bureaus were therefore treated as essential elements of the whole process of planning. By using numbers, the Russian Bolsheviks planned to modernize the whole of Russia which they considered backward. The planning and bureaucratization of this project through the use of various indicators was viewed as a way of promoting modernization and of overtaking developed countries. However, the gathering of information on each step of the diverse research and industry processes constituted a major burden. For instance, in the 1950s, preparing the annual plan for the Ural Machine Building Factory required the compilation of a document 17,000 pages long (Levine, 1967). Therefore, the Soviet Union tried to introduce mathematical and computer technologies as well as reporting networks in various branches of the economy. However, in practice, the process of gathering information for monitoring and central planning was very decentralized.

There were three types of scientific and technical work plans: annual plans (originally "control figures"), long-term plans (5–7 years), and perspective plans (15–20 years). The first type of plan was very detailed and, from today's point of view, can be understood as grant proposals for short projects. In such a proposal, the institution (not an individual scientist) provided the following information: the name of the institution, institutional partners, start and end dates, the content of the research and work stages, a list of principal investigators for each stage, the amount of financial resources necessary for the research and the sources of funding, a description of the expected results, and an approximate calculation of the assumed economic outcomes of the planned scientific work. This proposal structure does not differ significantly from today's grant applications. However, a key difference should not be overlooked: The topic of research had to be preapproved or, more accurately, ordered by central government authorities. Five-year plans contained fewer details (e.g., there was no division into particular stages of the work), while perspective plans set out the direction of research work on the issues considered by the central committees to be promising and timely.

The evaluation of plans within the framework of central planning can be divided into two phases. In the first one, the plans were designed, then evaluated, and accepted by the authorities. Approval by the authorities was not a form of peer review because there were no peers: The authorities were often party members and their expertise served primarily not only to evaluate the quality of planned research but also to assess whether the plan was in line with party goals. In the second phase, the realization of plans was reported on and verified. The evaluative aspect of the second phase of planning – essential from the perspective of the goals of this book – is practically ignored by most of the works on Soviet scientometricians as well as by English language studies on central planning in science.

With the example of Polish universities and research institutes, I now show how these two phases were implemented in practice. In the Polish People's Republic, as in the Soviet Union and other socialist countries, one of the key characteristics of science and higher education was the separation that was maintained between the higher education provided at universities and the research and development conducted at research institutes and institutes of the Polish Academy of Sciences. This does not imply – as is too often claimed in the literature – that there was no research at universities. Rather, it meant that universities and institutes constituted two distinct regimes that were governed by slightly different rules. Moreover, it should be noted that sovietization in Central and Eastern countries assumed varied faces, thus universities and research institutions, for example, in the Czech Republic, Poland, and East Germany, were transformed according to their specific, local contexts. One of these, for instance, was the fact that the percentage of professors who survived World War II influenced the speed of ideologizing science (Connelly, 2000).

For socialist decision makers, central planning was neither an administrative nor formal activity, but a purely scientific and creative task (Schaff, 1956). The planning of research work for universities was based on the coordination of the activities of departments representing a given scientific discipline. The plans of these particular departments were evaluated at the national level by 46 teams consisting of 193 members in total of whom 164 were university professors. Each of these teams determined what the key tasks of a given scientific discipline were and then evaluated the overall achievements of the discipline in Poland in relation to these key tasks. The last phase of planning was the evaluation and prioritization of the topics submitted by departments. The teams often corrected the submitted plans through direct discussion with the heads of department (Schaff, 1956). In the case of research institutes, there were three key sources of planning: the guidelines of the national authorities resulting from national economic plans, declarations by the central management of production facilities, and the initiative of the staff of these institutes themselves. Further, these plans had to be approved by the competent ministers responsible for a given economic sector.

During the first phase of centrally planned research, evaluation was based on the probability that the plan being implemented was accurate. In the case of basic research, assessments were also made as to whether research might also serve some other applied use. In addition, the quality of the planned research was compared with that of similar research in other countries (Tuszko & Chaskielewicz, 1968). Moreover, as in the Soviet Union, plans were evaluated to determine the most efficient means of carrying out tasks in light of the priorities outlined by the authorities. Beyond economic profitability, the criteria for plan evaluation included originality and the length of time the new technology might be used for, its safety and conformity with social needs (Nolting, 1978).

Evaluation within the second phase of centrally planned research was based on assessment of the research that had been performed. The results of evaluation were expressed both in terms of metrics (the percentage at which the plan had been implemented) and in a qualitative way by describing the novelty of the discovered phenomena and facts, the significance of research results, and the research's resonance within world literature. However, it was the metrics that determined the assessment result, and at most, qualitative evaluation only complemented them. The key criterion within the two phases was economic efficiency understood as the ratio of the results of scientific work to the expenditure (labor, materials) used in it. However, it was emphasized that in the case of basic research, effectiveness was translated into economic efficiency only after a period of time. Important to note here is the fact that publications were one of the key criteria of evaluation: The number of publications was counted, and publications themselves were measured in terms of their length as expressed in author sheets. Furthermore, during the 1950s and 1960s, the counting of citations was introduced as a good practice for assessment.

The critique of the use of publication length as a criterion for determining scientists' effectiveness was already made by Bukharin as far back as 1931, during the First All-Union Conference on the Planning of Scientific-Research Work. Arguing that this was a crude form of measure, he also stated that: "Here it is necessary to change to a system of complex indicators taking into account qualitative characteristics, or to indirect methods of *economic* evaluation (...) We still need a systematic elaboration of this problem" (as cited in Graham, 1964, p. 145). Almost a hundred years later, the Global North, South, and East are still faced with the same problem of determining the best qualitative and quantitative indicators for evaluating research and academic work. Moreover, as evidenced by both Bukharin's words and the broader picture of Soviet planning in science that I have been describing, the economization of academia today is not only a product of the neoliberal market that was introduced into science and higher education. For decades prior to this, research in Russia or Poland had to be economically efficient and the work of researchers was evaluated in light of how their outcomes contributed to the economy and society (within the prism of various plans) as well as in relation to the number of printed pages. Thus the economization and metricization of academia in evaluation systems does not only have a Western face. Its Eastern face – manifest today in the countries of the Global East – is its big sister.

Evaluation in central planning was ex ante evaluation because decisions regarding evaluation and incentives (bonuses when the plan was realized) were taken predominantly on the basis of estimates of return and not on actual results (Cocka, 1980). The degree to which a plan had been realized had to be assessed on an annual basis, and this was a task conducted within the second phase of planning. Although in Russia, as Dobrov (1969b) notes of the year 1965, the Soviet Academy of Science conducted 508 experimental and constructional projects of which 469 were fully completed, in Poland the implementation of central planning and the evaluation of its results appeared to be a demanding task. In the 1950s, only part of plans were completed (Schaff, 1956, p. 80), and the key measure of a plan's execution was the percentage at which it was implemented. Those involved in planning were aware that plans needed to be realistic, not least because of the fact that the meeting or missing of plan targets affected not only research institutions' incomes but also managers' careers. Thus, the level of plan execution was perceived as an indicator of general managerial capability just as, decades later, external funding became a criterion for the evaluation of university managers. The systematic qualitative and quantitative evaluation of a plan's execution was suggested as the core tool for improving outcomes. Such evaluation was supposed to be carried out by a department or institute's research council.

The central planning system had a number of achievements to its credit, starting from the production of the world's first earth satellite. Nonetheless, science planning compelled researchers and managers of research institutions to find ways of getting around the system, which is to say, of playing the evaluation game, finding back doors thanks to which regular work could be done. Playing the game was driven by the central level of planning and the shortage economy (Hare, 1989) which constituted the context for planning and implementation. Bad planning had varied repercussions including poor diffusion of scientific advances, a lack of coordination in many projects, and a general lag in technology development and use as compared with Western countries (Nolting, 1978). One of the most widespread problems, which stretched beyond the science sector, was the inadequate provision of the components necessary for research (e.g., materials, chemical substances). Thus, managers hoarded such materials in order to minimize the effects of supply failures (Ellman, 2014). In the list of requirements that were submitted annually, a much larger quantity was indicated for the tasks to be carried out (i.e. to produce 100% of the plan), because it was assumed that anyway, this quantity would either be reduced or not delivered at all. Thus, there was a shortage of material in some organizations and a surplus in others. Josephson's historical account (1992) of the early writing of five-year research plans by scientists (around 1929) illustrates another practical adaptation to new evaluation regimes. Initially, scientists presented detailed documents that were hundreds of pages long, but they eventually realized that such long documents were less likely to be examined as officials were much more likely to accept more general documents consisting of several dozen pages.

One can understand the attempt to get around the system of central planning as an evaluation game, that is a practice of doing science and managing research in a context transformed by the central authorities. The central bodies responsible for planning introduced new rules that included an obligation to provide detailed plans of research phases and of criteria for researchers' performance evaluation based on the length of their publications. Researchers and managers adapted to this situation very quickly, and a pronounced tendency to propose safe research problems that could easily be solved could be observed. On this trend, Kolotyrkin commented: "An institute can fulfill its subject and financial plans year after year without contributing anything to technical progress" (as cited in Cocka, 1980, p. 235). This was one of strategies deployed for following central regulations at the lowest possible cost.

3.3 Two Scientometrics Programs

Any history of Russian scientometrics must confront two core paradoxes. The first relates to the fact of treating the scientometrics born in the 1950s as Soviet or "Red" science when one of its two key figures, V.V. Nalimov, was imprisoned for several years in the Gulag and himself used imperialist science (cybernetics) to develop his program of scientometrics. The second concerns the view that Soviet scientometrics originated as a tool designed to control and plan science. In fact the opposite was true: Nalimov saw science as a self-organized organism. He based his reflections on cybernetics, which at the time was trying to redefine the understanding of control (steering) and free it from the language of hierarchy and coercion to give it a new meaning in the context of adaptation to complex systems (i.e., how to control a complex system to ensure its stability). We must also note here that the other key figure in this history, Gennady Dobrov, believed that scientometrics should be used to control and plan science. In the final analysis, however, as Wouters (1999) contends, it is highly questionable whether scientometrics had any impact on science policy in the Soviet Union at all.

In 2019, Glänzel, Moed, Schmoch, and Thelwall – four key researchers in today's scientometrics and bibliometrics community – edited *The Handbook of Science and Technology Indicators* (Glänzel et al., 2019). According to its editors, the book presents writing on state-of-the-art indicators, methods and models in science and technology systems. The handbook also includes a "short history of scientometrics" in which place is made – as it is in other handbooks for the bibliometrics and scientometrics community – for, among others, Derek de Solla Price, Robert Merton, and Eugene Garfield as founding fathers of the quantitative study of science. In the abridged history presented in this influential handbook, Russian scholars do not even receive a passing mention.

Even when Nalimov is mentioned in handbooks, however, it is only to note that he coined the term "scientometrics" (Andrés, 2009; Gingras, 2014; Vinkler, 2010). This is despite the fact that in 1987 he was awarded the Derek John de Solla Price Award, which is the most prestigious award in the scientometrics community for distinguished contributions to the field. One can nonetheless cite a few instances in which the Soviet contribution receives some recognition. For instance, in a book edited by Cronin and Sugimoto (2014), De Bellis states that "two influential scientometric schools had been established during 'Khrushchev's Thaw' by Gennady Dobrov in Kiev and Vassily Nalimov in Moscow" (De Bellis, 2014). Even as De Bellis acknowledges the Soviet contribution (De Bellis, 2009) however, his acknowledgment is mediated by a "Western researcher" whose work implicitly demonstrates the usefulness of Soviet research: "a key role in channeling Eastern 'Red' notions of science policy into the Western provinces was played by the British scientist John Desmond Bernal" (De Bellis, 2014, pp. 31–32). There are two major problems with presenting the history of scientometrics in this way: first, it reduces Nalimov's contribution to merely coining the term while his original program – inspired by Derek de Solla Price but also substantially different from Price's ideas – is overlooked. Second, the assumption that every work that appeared in the Soviet Union was "Red science," that is, marked by the Soviet ideology based on Marxism, Leninism, and Stalinism, is far too much of a generalization and in the case of Nalimov's program – as Wouters (1999) argues in his *Citation Culture* – simply incorrect.

The Soviet organization of scientific labor which began to be rolled out at the turn of the twentieth century and Soviet scientometrics, born in the 1950s, constitute two parallel programs for the development and modernization of science and research through varied metrics and policy tools. They are, however, distinct both on an ontological, that is what constitutes science, and an epistemic level, that is, how can science be measured and supported by planning. Moreover, there was not one but two schools of Soviet scientometrics: the first one founded by Dobrov in Kiev and the other one started by Nalimov in Moscow. Not only were the school of thought different, so too were their founding fathers: Dobrov was a party man, whereas Nalimov was imprisoned in a Gulag (Granovsky, 2001; Wouters, 1999).

Nalimov developed scientometrics as part of a "science of science" program (Skalska-Zlat, 2001; Wouters, 1999) that was actually initiated in the 1920s and 1930s (Kokowski, 2015, 2016), among others by Polish sociologists: Florian Znaniecki (1925) and later by Maria Ossowska and Stanisław Ossowski (1935) whose article "Nauka o Nauce" appeared in English in 1964 as "Science of Science" (Ossowska & Ossowski, 1964). In the Soviet Union at the same time, science studies [naukovedenie] were being developed, which focused not only on the organization of science but also on the philosophical and historical reflection on science as such (Aronova, 2011). For Nalimov and Mulchenko (1969), science was a self-organizing system directed by information flows. Thus, quantitative analyses of science should serve to improve science itself, rather than serving instrumental goals. This approach was inspired by American cybernetics which in the Soviet Union of the mid-twentieth century was treated as pseudo-science and an "ideological weapon of imperialistic reaction" (Granovsky, 2001). Moreover, as Aronova (2021) shows the rise of scientometrics in the Soviet Union and the key role of Nalimov in it was coupled with a transnational endeavor in which Eugene Garfield tried to sell his key product (the Science Citation Index) in Soviet Union and to find out how the Soviet approach to automating the extraction of scientific information from a scholarly publication is different from the American approach.

The purpose of scientometrics and studies of science was perceived in a completely different way by Dobrov (1969a). For him, it was the work of revealing regularities in the development of science that served as the basis for evaluating and regulating science and technology as well as for informing science policy (Wouters, 1999). In contrast to Nalimov, Dobrov perceived science as an organism greatly influenced by science policy.

Nalimov started from the assumption that science was controlled by information flows and by invisible research collectives. He nonetheless stressed the necessity of conducting science policy at the central level in order to make strategic decisions on resource distribution and the desired direction of development. The main question relating to the governance of the sector of science was how to combine the self-organization of science with the central planning of research. Nalimov looked for inspiration to other larger technical, biological, and social systems, believing that over-regulated systems were rarely effective. According to him, the measurement of science could be useful not only for central governments but also for invisible research collectives in order to promote better self-organization. In addition, metrics and indicators could act as intermediaries between the central and self-organized levels of science by building legitimacy for political decisions. Scientometrics could allow the central government to know in advance whether a given direction of research would turn out to be a blind alley for scientific development. Nalimov did not highlight economic efficiency but rather the development of science itself. This was because for him, scientometrics was a tool for understanding science rather than for controlling it. However, Nalimov's voice was ignored both by decision makers in the Soviet Union and by parts of the scientometrics community.

In Poland, the idea of using scientometrics as a basis for controlling and assessing research at the national level was first envisioned during the late 1960s. In 1968, Tuszko and Chaskielewicz, members of the Polish Academy of Sciences, wrote: "there is a need to 'measure' and 'determine' in a comparable and useful way to analyze both the research effort (also the scientific potential) and the results of the research work (…) and thus the need to create a kind of 'scientometrics' just as we have hydrometry for measuring hydrological phenomena, econometrics – economic phenomena, etc. Scientometrics would enable the use of concepts and indicators related to the development of science in a possible unambiguous way" (Tuszko & Chaskielewicz, 1968, p. 294). Here, scientometrics was combined with central planning focused on the economic efficiency of research. This is then a clear example of the economization and metricization of academia decades before the birth of NPM.

Today, the forms of economization and metricization that one can observe in Australia, the Czech Republic, Finland, Italy, Poland, or the UK look very alike. However, in the Czech Republic and Poland, the economization of academia did not start at the end of the 1970s, nor did mass metricization begin with the introduction of university rankings at the beginning of the twenty-first century. At the national level, the history of these processes went back several decades.

3.4 Socialist and Neoliberal Models

Socialist research evaluation systems based on central planning, the evaluation of plan feasibility, and their execution appear to be different from evaluation systems currently operating in Anglo-Saxon and other countries of the Global North. Socialist systems were based on different goals and values, as well as other metrics and indicators.

The main difference is that while contemporary systems are based primarily on ex post evaluations and valorize research that fits into the model of research excellence, socialist systems deployed ex ante evaluation based on the approval and prioritization of research topics. This distinct approach derived from the assumption that it was possible to directly influence the development of science and to subordinate it to broader social goals.

I present key dimensions of these two models, the socialist and neoliberal, in Table 3.1. This table is focused on the national research evaluation systems system that are designed for institutional evaluation. Therefore, it indicates the dominant, not the only, type of evaluation. It is worth noting that also in evaluating individual researchers and their activities, ex-ante evaluation plays an important role, for instance, in applying for research funding. Nonetheless, it is not the dominant model of the relationship between the political institutions of the states and the state. I opted not to attach additional geographical characteristics to the names of models because these models cross over various classifications: the socialist model could be called "Eastern" while the neoliberal one is not (only) "Western." Further, as argued above, this East–West binary lost its purchase after the fall of the Iron Curtain. Today the neoliberal model is being implemented both in countries of the Global North (e.g., Australia and the UK) and of the Global East (the Czech Republic and Poland) which, half a century earlier, had followed and developed the socialist model.

The socialist model elaborated in Russia at the beginning of the twentieth century, developed by the Soviet Union and implemented in other socialist countries, was based on the premise that science is not an independent landscape but should serve the development of the socialist economy and socialist society. To achieve this vision, research, like other sectors of the economy, was incorporated within central planning and more specifically, in a framework of ex ante evaluation. The key method of evaluation was a verification of research plans in light of preapproved research themes and programs. The economization of the science sector was manifest in the view that science was a field subject to the economy and its metricization through metrics served not only to make decisions but also to control the realization of plans.

The neoliberal model, grounded on ideas connected with NPM, emerged in the 1980s. The first such model was introduced in 1986 in the UK and later in other European countries. In this model, there is no preapproval of themes and programs and the key function of evaluation is to rank evaluation units according to their performance and, in many countries, to use this ranking to determine funding distribution. Thus the economization of research is evident in the understanding of the research process as a linear economic input–output model on the basis of which metricization of the whole sector is implemented. In this context, being a good performer in terms of an evaluation unit does not mean achieving a target or realizing a plan but rather, being better than other units. Thus, targets are often not known.

Table 3.1 A comparison of socialist and neoliberal models of national research evaluation systems within research funding systems

Dimension of model	Socialist model	Neoliberal model
Founding period	From the1920s in the Soviet Union and the 1950s in Eastern Bloc countries.	From the 1980s.
Countries	The Soviet Union and other socialist states.	Australia, the Czech Republic, Denmark, Finland, Italy, Norway, Poland, the United Kingdom and others.
Status of model	Model not in use anymore although its heritage still plays a role in former socialist countries. The character of the Chinese research evaluation system is debatable.	Model is currently being used, developed, and implemented in various countries.
Dominant type of evaluation	Ex ante	Ex post
Primary function of evaluation	Verification of research plans in light of preapproved research themes and programs.	Ranking of units according to their performance and in many countries, the use of this ranking to determine funding.
Understanding of science function	Development of the socialist economy and society.	Organizing knowledge about the universe. Knowledge can be useful for the development of the economy.
Founding values of evaluation	Research is not an independent sector of the economy and has to serve it and socialist society (idea of the New Soviet Man). Therefore, research has to be in line with socialist state goals.	Investment in research as part of the public sector should be measured and research institutions must be held accountable for using public funds. Transparency of spending and assessment procedures is important and in line with ideas connected to New Public Management.
Economization of research	Economization of research through the perception of science as a field subjected to the economy.	Economization of research through the perception of the research process as a linear economic input–output model and the subordination of science to the economy.
Metricization of research	• Metrics are used as a key element in the assessment of the plan's execution: both on the level of plan itself (how much of the plan, expressed as a percentage, has been implemented?) as well as performance of researchers (how many pages were published?). • Metrics are not used to make decisions but to control the realization of plans. • Metrics are used to set targets within plans.	• Metrics are used to make decisions on the basis of rankings and comparisons of performance of evaluation units. • Metrics are used as targets although frequently, due to constant comparison across evaluation units, targets are not set and communicated. • Being a good performer does not imply achieving a target but rather being better than others.

Table 3.1 (cont.)

Dimension of model	Socialist model	Neoliberal model
Key tools of evaluation	• Planning (only topics consistent with the founding values were allowed). • Reporting (how well did the research conducted conform to the plan?). • Assessment of the plan's execution: panels and expert-based assessment by other researchers and Party members. • Widespread scientometrics analyses of institutions.	• Reporting: topics are not pre-approved although in some funding instruments research has to fit into predefined themes (e.g., grand societal challenges). • Bibliometrics indicators and peer review conducted by other researchers. • Assessment of the excellence of the research and its societal impact.
Unit of evaluation	Research theme or research program within a particular institution.	Research institution or its subparts, research group discipline within institutions.
Basis for assessing whether a result of evaluation is positive	Assessing whether planned research is in line with the themes and values indicated by authorities.	Performance-based in terms of research excellence, societal input, and economic efficiency.
Approach to societal impact of research	Each research project has and should have impact on socialist society.	Societal impact of research has been included in research evaluation systems only since the beginning of the twenty-first century.

My aim in relating this history of central planning in the Soviet Union, and the ex ante model of socialist evaluation has not been to argue about who was first in terms of introducing national research evaluation systems. Rather, in narrating this past, I wish to underline the point that certain systems – in this case the neoliberal ones – have been built on the ruins of previous (socialist) systems. Because it is socially embedded in culture, heritage, and traditions, one must remember that research evaluation does not take place in a vacuum. Thus, the fact that a new system was built on the remains of the old has had a substantial influence on how, in countries like the Czech Republic or Poland, researchers perceive research evaluation systems and how research institutions in these countries adapt to new evaluation regimes. Although evaluative powers might look similar across countries today, the context of their implementation can differ substantially, and this influences the ways in which researchers and institutions play the evaluation game

The transformation of socialist systems into neoliberal ones was not the product of an internal modification of the research evaluation system itself. Rather, it was an imposition brought about by a broader logic of transformation that encompassed whole countries and economies. In 1993, the American Association for the Advancement of Science, the Polish Society for the Advancement of Science and the Arts, and the State Committee for Scientific Research in Poland organized a workshop in Pułtusk (Poland) that was co-financed by the US National Science Foundation. Its theme was the use of peer review and quantitative techniques in the evaluation of science and scientists in East-Central Europe, in relation to experiences in Western Europe and the United States. The results of this workshop were published in a book entitled *Evaluating Science and Scientists: An East-West Dialogue on Research Evaluation in Post-Communist Europe* (Frankel & Cave, 1997). The book provides interesting evidence of how the shift to a market economy has led to reforms in research funding, management, and evaluation. Participants of this workshop recognized a number of major issues in the transition period. The first one was the international isolation and poor research quality created by centralized funding and management. The second was the implementation of quantitative or bibliometrics indicators and their combination with peer review. Thus, it appeared that peer review was perceived as a solution to all the problems of postcommunist countries. However participants of the workshop were aware that peer review constitutes a radical break with centralized funding because it gives scientists (and not the authorities or party members) a role in determining which research project should receive funding.

For the past three decades, no Soviet science organization has existed, although some elements of the official socialist approach can still be discovered, for instance, in China. Soviet planning has been replaced by a form of organization based mostly on the mix of a market economy and the strong institutional autonomy of academia. Foreign experts were invited to former socialist countries as part of the process of replacing the ruling logic. One should not underestimate their role in trying to convince people, through all available media, of the need to accept neoliberal shock therapy, the closure of workplaces, and the tightening of belts. The above-mentioned workshop in Pułtusk is just one example of the many ways in which neoliberalism was implanted on the grounds of a ruined socialist way.

The socialist model's heritage is still visible in various practices of day-to-day academic work, and the past means a lot for the present situation in the Global East. First, as Sokolov (2021) argues describing situation in Russia but it also applies to other post-socialist countries, the use of quantitative indicators is an expression of distrust toward scientists as evaluators. This heritage is also still tangible in former Eastern Bloc countries where trust in experts was substantially lower than in other societies and where metrics are perceived as tools for finding a way to get around experts and peers. Second, research planning of any form, for instance in the framework of grant applications, is occasionally perceived as limiting the autonomy of science and academia. Third, perceptions of the nature of science and research during the first days of the transition from the communist regime are still vital to this day. In 1990, Tadeusz Popłonkowski – Deputy Minister of Education responsible for higher education in Poland – said that during parliamentary work in 1990, the prevailing mood in the academic community was to give academia all the power and prevent public authorities and the government from interfering in any university affairs, including the spending of public money, tenders, and labor law (Antonowicz et al., 2020). Researchers and research institutions that abandoned the socialist model of research evaluation and funding wanted to be fully autonomous and independent. Finally, any form of evaluation – even when presented as a tool of accountability or for showing the utility of research to society – is perceived, first of all, as limiting the autonomy of research.

In post-socialist countries, adapting to research evaluation regimes, playing the evaluation game, and following evaluation rules at the lowest possible cost are often treated not as gaming or cheating but rather as a way of fighting with the system. Thus the fight against (evaluative) power and with anyone who controls, monitors, or assesses research is sometimes treated as virtuous. This attitude owes its vigor to a still vibrant heritage which in Poland derives from the 123 years of partitions and almost half a century of communist rule that its people were confronted with which. This heritage also expresses itself in approaches to research integrity and ethos in higher education. For instance, in former communist and socialist countries, cheating by students in exams is rarely reported on by other students because it is treated as a widely accepted practice (Hrabak et al., 2004; Teodorescu & Andrei, 2009). One can explain this by reference to those social and cultural factors specific to countries in the midst of postcommunist transitions to a market economy. Hence, playing the evaluation game, gaming the number of citations or other forms of adaptation to evaluation regimes may be treated (by other members of academia) as a form of commendable resistance.

I now turn to a consideration of those varied evaluative powers that can be identified today at the three levels of the glonacal perspective. This discussion lays the grounds for the subsequent investigation of practices around playing the evaluation game and their analysis in the context of the implementation of a given evaluative power.

4

The Diversity of Evaluative Powers

During a specially convened meeting on January 29, 1985, dons and senior administrators at Oxford University decided by 738 votes to 319 that Margaret Thatcher would become the first Oxford-educated prime minister not to be awarded an honorary doctorate. The assembled decided to withhold the honor because they deemed it "unreasonable to expect the University to honor the holder of even the most elevated office under the Crown, and to ignore the effects of her policies on the values and activities we are committed to as academics" (Hilton, 2013). Dr Denis Noble, one of those organizing opposition to granting the doctorate, commented after the vote that in this way, Oxford was protesting "against the damage inflicted by government policy on science, education and health" (Wainwright, 1985). The Thatcher administrations were perceived as inflicting profound and systematic damage to the public education and research system. The refusal to honor her was a way of showing both that honorary degrees ought not to be regarded as automatic entitlements, and that as an Oxford alumna, they believed she should better understand the needs of the higher education and science sectors.

When Thatcher became prime minister in 1979, she was confronted with a recession that laid the grounds for the emergence of the evaluative power of the state. Both the state's role and leverage were simultaneously strengthened by demands for accountability and weakened by mass privatization and the empowerment of free markets. This was the period during which Thatcher redefined relations across the state, public sector, and markets, aligning them with neoliberal precepts. This approach became the foundation of "Thatcherism," that is a form of government based on the promotion of low inflation, minimal government involvement in public policy and the private sector, and repression of the labor movement. This was also the time in which the set of ideas referred to as New Public Management, which was based in part on the idea of accountability in public funds, matured and flourished. Most of the controversies that marked the early years of Thatcherism erupted when cuts to public sector spending were announced, at a time in which the gross domestic product was falling and unemployment rising sharply (Burton, 2016). During this period, research funding, as distributed through the research councils, was kept level, although funding for universities was cut (Agar, 2019). In this context, the British government initiated the design of the first national research evaluation system which, as Agar highlights, was "largely driven by causes internal to science funding bodies rather than being driven by external demands for greater accountability and selectivity" (2019, p. 116). Thus the University Grants Committee, chaired by a mathematician, Peter Swinnerton-Dyer, was appointed for this task.

Swinnerton-Dyer introduced the Research Selectivity Exercise (RSE) – the forerunner of the Research Excellence Framework (REF) – in 1986. This was the first neoliberal national research evaluation system used to distribute funding to universities. Swinnerton-Dyer has since said that while he was responsible for the creation of the RSE, "it has been amended in ways to make it work less well since" (Swinnerton-Dyer & Major, 2001). He did not, furthermore, expect his tool to exert the kinds of influence it has had. Thus the initiator of the RSE considered the ultimate fruits of his work, developed over the years in accordance with the original assumptions, to be inferior to the original.

Behind the introduction of a new country-wide competitive regime was an effort to safeguard funding for universities in the face of the cost-cutting Thatcher government's drive to optimize the science and education budgets. Thus academia proposed its own way of changing the model of funding distribution. Swinnerton-Dyer wanted "to find a system for allocating money that was fair, but a system that was not egalitarian" (Swinnerton-Dyer & Major, 2001). This system was based on the assessment of academic papers published in prestigious scholarly journals and focused purely on basic research. Social impact – such an important criterion of the current version of the British system – although vital, was neglected as a factor for funding decisions. In 2007, David Eastwood, chief executive of the Higher Education Funding Council for England, argued that British science would not be where it was without the evidence provided by assessment exercises. According to him, "the RAE [a successor of RSE] has done more than drive research quality; it has been crucial to modernization" (Eastwood, 2007). Attempts were made to enrich this modernization tool by adding bibliometric indicators (including the Journal Impact Factor [JIF]). However, the British system has remained mostly expert-based, although its findings are communicated through simple metrics. The quality of work is expressed by the number of stars: from no stars (unclassified quality) to four stars (world-leading quality); these have become the metric-founded targets for researchers working in British academia. As with the idea launched by Swinnerton-Dyer, the integration of the

economization and metricization of science reaches its peak in national research
evaluation systems that serve to distribute funding.

* * *

For many scholars reflecting on the state of academia today, the use of metrics to manage and evaluate the science and higher education sectors first began
with Thatcher. As I showed in Chapter 3, however, this is in fact a much older
story. Nevertheless, it was the British system that reignited the use of metrics
to organize science. Since the inauguration of research evaluation in Britain,
many countries have implemented national research evaluation systems. In
some instances, they play a role in performance-based research funding systems.
National research evaluation systems designed for higher education institutions
operate in, among others, Australia, the Czech Republic, Finland, Denmark,
Italy, Norway, and Poland. Some countries such as Italy, Poland, and Spain,
also have long-running centralized national systems designed to assess individual researchers during academic promotion procedures. A number of overviews
and typologies on national evaluation procedures are available (Geuna & Martin,
2003; Hicks, 2012; Reale et al., 2017; Zacharewicz et al., 2019). These classifications focus mostly on key methods of evaluation (peer review versus bibliometrics) and the implications for funding of evaluation results. Ochsner et al.
(2018) show that even research evaluation experts from the same country often
cannot agree – because of the high level of complexity of such systems – whether
the key evaluation method in their national system should be described as, for
instance, metrics-based or informed peer-review. This only underlines the fact
that presenting almost a dozen such systems in a single chapter is a challenging
task that should be undertaken in accordance with clear criteria which I lay out
below. In this chapter, my aim is to present an overview of research evaluation
systems whose key pillars are well described in the literature. More important,
however, is the fact that I focus on those systems whose evaluation games I
explore in Chapter 5.

Research evaluation systems are technologies of evaluative power that are, to
varying degrees, driven by metricization and economization. When one thinks
of these two forces within academic work (described in Chapters 2 and 3), as
two dimensions of any given system, then one sees that some systems are more
strongly metricized or economized than others. In this chapter, I characterize
research evaluation systems through the lenses of these forces. The degree of
metricization of a given system will be determined by both the extent to which
various metrics and bibliometric indicators (most often produced by global actors
like the Clarivate Analytics or Elsevier) are used within it and by the manner in
which these metrics are used (e.g. whether they serve only to monitor or also to

control institutions; whether they are used for the distribution of funds and/or as financial incentives within systems). The degree of economization of a given system is determined both by defining the effects of scientific work treated as products (mostly publications) and by determining the productivity level of scientists or institutions, as well as by looking at whether the results produced by a given research evaluation system have substantial economic consequences (for institutions or individual researchers).

In this chapter, in order to provide the context for Chapter 6's analysis of diverse forms of evaluation game, I survey the contemporary landscape of technologies of evaluative power, tracing them across three intersecting planes (the local, national, and global). Thus, this chapter sheds light on those powers that have shaped academia today. Starting with global evaluative powers, I focus on Impact Factor, international citation indexes, and university rankings. Although these powers are global, following the glonacal heuristic, I argue that they also play an important role at the national and local planes. Thus references to these global powers are inevitable in the elucidation of national evaluative powers (both monetary rewards systems and research evaluation regimes within performance-based research funding systems). Finally, I present two examples of evaluative powers (the centralized academic promotion system in Poland and monetary rewards systems in various countries) designed at the national level for national and local use, but which differ significantly from the previously discussed systems in that they were designed from the outset for the assessment of individual researchers.

Evaluative powers are global when they address institutions and people not limited to one specific country and when they are not shaped by the policies of a single country. However, some evaluative powers designed for national and institutional purposes, like the REF in the UK or the Norwegian Model can also influence global discourse and serve as a point of reference or inspiration for other countries.

In this chapter, I characterize the technologies of evaluative power and their levels of metricization and economization at each of the three intersecting planes which most influence the forms of evaluative game, and which can provide a comprehensive view of current research evaluation regimes in the Global North, South, and East. Moreover, using institutional cases, I illustrate how global and national forces produce and sustain their power through monetary reward or tenure track systems.

In discussing these systems, I sketch a general overview of the context of implementation and of the main elements of each system. This chapter does not attempt a comprehensive presentation of the past and present of the systems (references to key texts are provided to readers). Rather, by focusing on the key factors at play,

I show how strongly a given system is metricized and economized. In other words, the chapter points to the ways in which evaluative powers use economization and metrics both as tools of modernization and as means of controlling academia.

4.1 The Evaluative Power of Global Research

The account of the evaluative power of global research that follows is focused on global phenomena like the JIF. In addition, it considers institutions, for instance, when they promote international citation indexes instead of mechanisms that are internal to science or research itself like peer review, which is widely used but also decentralized. Numerous actors can be included within a history of global powers. For example, the US Government which in the 1920s launched the first science and technology statistics in the West (Godin, 2005), or, going further back, Imperial Russia and the Russian Bolsheviks who used statistics and numbers to organize scientific labor In addition, during the second half of the twentieth century, institutions like the European Commission, the OECD, UNESCO, and the World Bank have played an important role in shaping the science, technology, and higher education landscapes by implementing metricization and economization as modernization tools. In this chapter, however, I focus on three global technologies of evaluative power which have had an indisputable and direct impact on current research practices and which have shaped the evaluation game across the whole of academia. Moreover, these three technologies, that is the JIF, international citation indexes like WoS and Scopus, and university rankings, are most often identified by researchers in daily discussions not only as *tools* of metricization and economization but also as their main *source*.

4.1.1 The Curse of the Journal Impact Factor

Without a doubt, the JIF is the most well-known and widely used bibliometric indicator. It is frequently perceived as a symbol of research evaluation and as the main expression of its abuse (Davis, 2016; Else, 2019). This particular technology of power is highly metricized (one could even say that the JIF is a yardstick in science metricization) and economized (it widely used as both a metric and an incentive tool in various research evaluation systems). Originally developed as part of the creation of the Science Citation Index, and in order to help libraries make purchasing decisions, the JIF has become a hallmark of current ways of thinking about productivity and scholarly communication in academia. Some policy makers and researchers believe that articles published in journals with a JIF of 292.278 are of much greater worth than papers published in journals whose JIF is 1.838, and far superior to articles published by journals without a JIF rating. Although the highest ranking journal in 2020 has a JIF rating of 292.278, half of the journals indexed in the JCR 2019 have an impact factor below 1.838.

The JIF is widely used in academic review, promotion, and tenure evaluations at the institutional and national levels. In the United States and Canada, 40% of research-intensive institutions mention the JIF in their assessment regulations (McKiernan et al., 2019). More interesting is the fact that researchers use the JIF to present the value of their research in academic promotion procedures, including in the humanities, even when is not required (Kulczycki, 2019). One can chart varied patterns of use that are generated as a result of different national and institutional policies. Hammarfelt and Haddow (2018) have compared the use of bibliometric indicators among humanities scholars in Australia and Sweden: 62% from the former and 14% from the latter use the indicators even though many of them are well aware of the fact that metrics like the JIF do not accurately reflect publication patterns and traditional evaluation practices in their field. This is particularly noteworthy as the JIF is not calculated for humanities journals, for which there exists a distinct database named the "Arts & Humanities Citation Index" which lists journals but does not provide bibliometric indicators. This use of the JIF and other metrics can be understood as a form of "metric-wise" behavior (Rousseau & Rousseau, 2017), that is, as skills in using scientometric indicators to present the value of one's research. However, the fact of being metric-wise can reduce the intrinsic motivational factors for doing research.

Years after he founded the JIF, Eugene Garfield compared his tool with nuclear energy – a mixed blessing which, in the right hands could be used constructively, but in the wrong ones, could be subject to abuse (Garfield, 2005). Garfield had introduced the concept of the Science Citation Index (Garfield, 1955), later co-creating the JIF in early 1960, and publishing its first calculations in 1975.

Trained in chemistry, Garfield aspired to come up with a more efficient way for scientists to find relevant papers in specific fields, as well as a better mechanism for circulating critique of problematic papers. These were the motivations behind the Science Citation Index, with the JIF created as a means of selecting journals for it (Garfield, 2005). Although this tool was useful for the library and information studies community, in terms of collection management, researchers also quickly understood the potential consequences of using the impact factor outside of its original context. Just after publishing the first edition of the Journal Citation Reports with calculated impact factors for journals, Hanley (1975) wrote in a letter to *Science* that this metric created problematic incentives in that it prompted researchers to focus on amassing more citations than their scientific rivals. Today we are faced with the repercussions of almost fifty years of the evaluative power of the JIF. Conceived of as a tool to support researchers, the JIF was rapidly transformed into an apparatus for the metricization and economization of science.

The idea behind the JIF is simple. Garfield aimed to create an association-of-ideas index based on citations from journal articles. To help select additional source journals, Irving H. Sher and Garfield reorganized the author citation index (originally intended as an association-of-ideas index) into the journal citation index (Garfield, 2006). The definition of the JIF is as follows: The JIF is the yearly average of citations that papers published in a specific journal received over the preceding two years. This simple definition is what makes the JIF easy to use.

However, as I showed in Chapter 3, this ease of use is in reality due to the fact that almost all the diverse modalities (i.e., signaled limitations of any metric and calls for contextualization each time a given metric is used) of this bibliometric indicator have been dropped. Most of the time, JIF users are not aware of the distinction between citable and non-citable items (e.g. editorials that are not counted in the JIF denominator), which substantially influence the value of the JIF (Davis, 2016). Many are also not aware of the skewness of citations (a small number of papers gets the lion's share of all citations), a share of journal self-citations or even that there is also five-year impact factor. Further, researchers for the most part do not know that only citations indexed in Clarivate Analytics (current owner of the WoS) products are used for calculating the JIF which is why citations of their work in scholarly books are not likely to be counted (Kulczycki et al., 2018). Without these modalities, the JIF's value gets de-contextualized, which is how it becomes possible to use it in varied ways, for example, to summarize JIFs or to compare journals from different fields. It is important to ask why modalities are crucial for understanding the global power concentrated in Clarivate Analytics.

The story about the problems created by the JIF has been told many times, in different fields and from varied perspectives (Bordons et al., 2002; Bornmann & Williams, 2017; Else, 2019). Critiques focus on multiple issues, such as the non-exclusion of self-citations, the inflation of the JIF through inclusion of review articles and long articles, and the preference for publication in English which skews JIF indicators in relation to publications in other languages. The list of problems associated with the JIF is thus long and well documented, both in the scientometrics and bibliometrics communities, as well as in policy papers (Bonev, 2009; Elsaie & Kammer, 2009; Hicks et al., 2015). The key target of criticism is the assumption – made by various funding agencies, tenure committees, and institutions responsible for national research evaluation – that a journal is representative of its articles. It is on the basis of this assumption that the most relevant the apparently most cogent argument is provided by proponents of the JIF. It goes as follows: Calculations based on citations assigned to a journal can say something about the quality of an article published in that journal and, in the consequence, the sum of such calculations – that is a sum of the JIFs of journals in which a given researcher published – provide information about the quality and productivity of a given researcher. Naturally, as should be evident to almost anyone, this leap from a journal indicator to an article and ultimately an author indicator, what is called the Author Impact Factor or the Total Impact Factor (Kulczycki, 2019; Pan & Fortunato, 2014), is not justified. This fact is obvious to researchers even without expertise on bibliometrics because they know that the number of citations their paper receives depends on various factors (Bornmann & Daniel, 2008), with researchers citing or not citing others' work for many different reasons (Cronin, 2015).

The evaluative power of the JIF is most manifest in its use in creating lists of "top-tier journals." It is understood that to publish in top-tier journals implies that one is reaching a wide and relevant audience, which allows one not only to communicate research results but which can also help advance research career milestones. According to Bradford's (1948) law of scientific journals, journals from across scientific fields can be divided into three groups, each with about one-third of all articles published in a given field. The first group consists of the smallest number of journals which nonetheless, like the other groups, publish one-third of all articles. Garfield claimed that "a small number of journals accounts for the bulk of significant scientific results" (Garfield, 1996). Given that in order to get an overview of a field it might suffice to follow only these journals, they can be called "core" or "top-tier" journals. Policy makers use the concept of "top-tier" journals to introduce diverse evaluation and incentive instruments which result in the economization of this field. For example, these include national lists of scholarly journals used in performance-based research funding systems (Pölönen et al., 2020), national lists of journals used as monetary reward systems for publishing in a defined set of top-tier journals (Quan et al., 2017), and national lists of scholarly journals used in academic promotion procedures (Else, 2019). These national-level technologies of evaluative power will be addressed in the next section.

As a metric for analyzing journals, the JIF is effective. Thus, the challenge is not how to create a better indicator for assessing journals (dozens of other indicators like the Source Normalized Impact per Paper or Eigenfactor are available), but how to deal with the fact that the indicator designed for journals is used to evaluate individual researchers and their papers. Where the JIF is used to evaluate institutions, arguments about the mismatch between the JIF and the unit of assessment (i.e., institutions) are not common even though it has been shown that, for instance, the results of the REF2014 in the UK could be reproduced only to a limited extent through metrics (Higher Education Funding Council for England, 2015). This means that peers' contributions to the final assessment were significant, a fact that has also been observed in other evaluation exercises, for instance in the preparation of national journal rankings, in which assessments made by peers are compared with metric-based evaluations (Kulczycki & Rozkosz, 2017).

It is the ambiguity of the JIF as an indicator that is unsuitable for the assessment of individual researchers but adequate enough for the evaluation of journals or institutions, that prompts researchers to make frequent use of it in the course of daily practice. Thus despite ongoing critique of the JIF, they may call it up when they have to choose a scholarly publication channel for publications, or when they have only a few days in which to assess dozens of post doc applications. My experience in various evaluation committees has shown me that only a small share of senior researchers is fully aware of the modalities influencing the value of the JIF: too often they are convinced that all of science looks and works in the same way as their own discipline. From the optic of producing a more fair and balanced system of evaluation, the best strategy is not to argue against the use of the JIF altogether, but rather, to highlight the distinct modalities at play. Thus one should stress that discipline can influence the JIF value and that in the case of humanities and social sciences, an important component of citations is not taken into account at all because scholarly book publications are barely indexed at all.

In sum, the JIF has been internalized by many scientists and has today become a measure of science. While dropping it would be possible, academia would then need to search for another measure.

The JIF, as one of the key metrics used in research evaluation (many even consider it *the* metric) is evaluative power in a highly metricized form. It is, furthermore, intensely economized in that through it, diverse global, national, and local actors reduce research practices and scholarly communication to numbers that represent publications as the key product of research.

4.1.2 International Citation Indexes

Without the Science Citation Index, there would be no JIF. As noted earlier, through the former, Garfield aimed to create an association-of-ideas index. After over a half century, the scientific information and bibliometric indicators industry constitutes one of most widespread and yet hidden of evaluative powers. It is significantly metricized (along with raw bibliographic data, numerous indicators are provided) but moderately economized (it is used as a data source in various systems and university rankings, but not as extensively as the JIF). Garfield's work contributed to the creation of international citation indexes like the WoS Core Collection (WoS) and Scopus. Such indexes, owned by private companies, are one of the key drivers of global academia today and crucial technologies in research evaluation. Global actors like the World Bank, OECD, or UNESCO operate on the supranational plan, provide reports or even influence the higher education and science sectors through the provision of various classifications like the OECD fields of science and technology (OECD, 2015). Yet it is the content of citation indexes that actually shapes the daily practices of researchers (the literature review and choice of publication venue), managers, and policy makers (the benchmarking of institutions and countries, and comparison of results from university rankings).

WoS and Scopus are products that provide raw bibliographic data and bibliometric indicators based mostly on citations of scholarly publications indexed in those databases. These databases differ in the number and type of publications they index, the bibliometric indicators used, and many other details (more precisely: modalities). However, from the perspective of research evaluators, they are viewed as largely objective sources because they are based on numbers and not produced by the institutions conducting evaluation. Yet the rectors, deans, policy makers, and journal editors who are the consumers of these databases are mainly non-bibliometricians who rarely know how to analyze raw data. Their interest is in obtaining fast and clear results that show the position of a given country, institution, journal, or researcher in relation to their competitors. Thus benchmarking platforms (InCites for WoS and SciVal for Scopus) are provided for both databases. In this way, with just a few clicks, anyone can visualize research performance, develop analyses of collaborative partnerships, or explore research trends. Every interested user can access plenty of visualizations, charts, tables, and numbers. These come without context or modalities, just very clear results showing, for example, a rank of institutions.

Web of Science Core Collection (its web-based version was launched in 1997) and Scopus (established in 2004) are competing world-leading citation databases. Both are subscription based which is important in light of the global use of these products. On the global market, there are other similar products like Google Scholar, Microsoft Academic, or Dimensions (Visser et al., 2021). However, because of their scope, strict journal selection criteria, data quality (established by WoS and Scopus themselves) and power relations, the competitors of WoS and Scopus are not as popular. WoS's origins lie with the Institute for Scientific Information in Philadelphia founded by Garfield in 1956, which was in 1992 acquired by Thomson Scientific & Healthcare (later part of Thomson Reuters Corporation), and then in 2016 by Clarivate Analytics. Scopus is a product of Elsevier, a Dutch publishing and analytics company which is part of RELX corporate group.

It is critical that we highlight the ownership structure of these bodies – academia and science are today financed mostly by public funds but driven by citation indexes maintained by private companies. Although the formulas for calculating bibliometric indicators like the JIF are publicly available, the results of calculations depend on a number of factors. These are the degree of coverage of the citation index (e.g., which scholarly publication channels are indexed and which are not), how particular papers are classified as eligible/ineligible for inclusion, and finally what the threshold is for self-citations in given journal in order to exclude it from a citation report. These are the modalities that researchers in the research evaluation and bibliometrics fields are aware of, but detailed information is not accessible to them. Citation indexes are black boxes (Gingras, 1995; Wilson, 2016): everyone knows what data are included, yet few know why certain data are not indexed and how the data are classified. Out of this black box, a product is confected and expressed as a number, e.g., a JIF of 2.059. All business and computing processes are compacted into this number. It is not to the advantage of a private company to uncover the network of interests and power relations that produce this one number. It should be in the interest of public institutions to do so. The matter is, however, complex, and public institutions actively cultivate their dependence on data produced by external bodies. In this way, they can highlight the objectivity of indicators and figures and the fact that they themselves have had no influence on the method of calculation or on the calculation itself.

It should be noted that both the way in which indicators are constructed and the range of indexed sources (i.e. whether non-English journals are sufficiently represented) have a significant impact on the picture of science that is disseminated both at the global level and in the countries concerned. For instance, a comparison of bibliographic databases of scholarly publication in social sciences and humanities in eight European countries (Kulczycki et al., 2018) reveals that in some countries (e.g. Poland and Slovakia), less than 16% of publications in 2014 are covered in WoS, while even for one of the most internationalized countries like Denmark, WoS covers only 51% of publications. The consequences are not only a distorted picture of publishing practices in various fields and countries but, more significantly, the setting of targets based on incorrect data. This matters because data from WoS and Scopus are used, among other things, as a basis for devising the most influential university rankings, which are key reference points for most countries and institutions, and through the results of which decisions regarding the distribution of funds are made.

International citation indexes are then highly metricized and economized, although less so than the JIF because one of their ongoing aims is to provide raw bibliographic data which can be used to build metrics or rankings as part of different evaluation regimes. Like the JIF, this evaluative power plays an important role at the global, national, and local levels where, as an "objective source," it is used to produce indicators and support decisions.

4.1.3 Global University Rankings

While the JIF and citation indexes are critical evaluative powers for scholarly communication, they are largely invisible to the general public. The most widely used and known global evaluative powers that contribute to the metricization and economization of the sector are global university rankings like the Shanghai Ranking (ARWU) published since 2003, the QS World University Rankings published since 2014 by Quacquarelli Symonds in collaboration with *Times Higher Education,* which started its own World University Rankings in 2010. These rankings are extremely metricized (almost as much as the JIF) and economized (indirectly rather than directly, treating ranking positions as hallmarks of university quality that might provide various financial benefits). According to Niancai Liu, the founder of ARWU, Shanghai Ranking was "an unexpected outcome when we focused on uncovering the distance between us (Chinese elite universities) and the world-renowned universities (according to the scientometric scale of knowledge production competitiveness)" (Niancai Liu quoted in Luo, 2013, p. 167). The QS rankings were also created to "serve students and their families" (Sharma, 2010). Nevertheless, rankings have occupied the attention of global and national media as well as policy makers and managers in academia who use them to inform and justify the setting of goals and the provision of incentives to researchers and institutions.

The history of university rankings goes back to the United States in the early 1980s, when *U.S. News & World Report* published its first "America's Best Colleges" report. However, one could consider a number of earlier attempts, some dating back to the beginning of the twentieth century, as the first efforts to rank academic institutions (Hazelkorn, 2015). Wilbers and Brankovic (2021) argue that the advent of the university rankings was possible by the rise of functionalism to the status of dominant intellectual paradigm what has made universities performing entities. While students and their parents have been the primary audience for these rankings, other stakeholders, especially governments, ranked higher education institutions, the public and media also pay attention to the ranking results and to how institutions from their countries or sector move through the ranks. Key drivers in the production of global rankings are the economization of the higher education and science sectors resulting from the transition to knowledge-intensive economies, and the global pursuit of talented students (Espeland & Sauder, 2007; Vernon et al., 2018).

Rankings results are differentiated and varied by criteria, weighting, and the way in which data are collected. Universities are assessed in terms of the quality of their teaching and research but also their level of industry income and academic reputation as reflected in surveys of other universities. Rankings are based mostly on institutional data that are problematic to access and use because of its sensitivity as well as because of the diversity of reporting practices. Thus most ranking institutions strongly rely on bibliometric data. In this way, the global evaluative power of rankings goes along with the evaluative power of citation indexes and the JIF. *Times Higher Education*'s announcement that henceforth, it would be using Scopus data instead of WoS, was an important piece of business communication for Elsevier (Scopus, 2014). Bibliometric data are easy to obtain but the heterogeneity of ranked universities makes the emphasis on modalities even more important for understanding the ranking process. However, the way in which bibliometric indicators are used to create rankings is dubious not only for ranked institutions but also for a section of bibliometrics experts.

The popularity of rankings rests largely on the ways in which their results are presented. Although there are university rankings in which no rigid table exists and each user can select the desired parameters (e.g. CWTS Leiden Ranking, U-Multirank), the popular imagination is governed by rankings in which a university or country's position is expressed by a single number: "our university is at 121," "we already have three universities in the top 100" and so on. To have at least one university in the top 100 institutions in the most popular rankings (like AWRU or THE) is often a policy goal that can transform an entire country's higher education landscape. An example was Project 5–100 in Russia which aimed not only to improve the rankings of Russian universities but also to increase the number of citations Russian researchers generate (Kliucharev & Neverov, 2018). At this point, it should be stressed that the choice of criteria and the source for bibliometric data substantially influence the results of rankings. As Moed (2017) shows, there is no such thing as a set of "the" top 100 universities in terms of excellence. Only 35 of the same institutions appear in the top 100 of 5 of the most popular rankings. Moreover, one can detect clear geographical bias in each of the rankings that position themselves as "global." U-Multirank is oriented toward Europe, AWRU toward North America, CWTS Leiden Ranking toward emerging Asian countries, and QS and THE toward Anglo-Saxon countries (Moed, 2017).

University rankings are a different form of evaluative power than that of typical research evaluation systems. First, most popular rankings are global even if there are numerous national versions. By contrast, research evaluation systems are implemented mostly on the national level. Second, the results of university rankings do not serve directly to distribute funding because the ranking institutions – mostly private ones – create rankings to make money, not distribute it. In the case of research evaluation systems that are part of national performance-based research funding systems, the results of evaluation are translated into money that is then transferred to evaluated institutions. Finally, in relation to this book's focus, the most significant difference is that university rankings do not motivate specific researchers to target their work toward improving the ranking of their institution. This is because an individual researcher's contribution to the university's final rank is difficult to discern, and in this, the ranking also differs from many national systems in which a specific number of publications is required of each researcher during the evaluation period. This does not hold where a researcher is a Nobel

Prize winner (Meho, 2020) who can substantially improve the rank of the institution employing him or her.

The pressures generated by university rankings are powerful incentives for universities to adjust their activities and to start playing the evaluation game. Using Foucault's concept of discipline, Sauder and Espeland (2009) investigate organizational responses to the ranking of higher education institutions. Their study shows that researchers and managers internalize rankings pressures and reacted to them by means of varied forms of symbolic response such as decoupling and gaming. In their design, rankings reduce the heterogeneity of universities to a handful of numbers and show what economic inputs provide the best rankings that can translate into profits (endowments, students fees, etc.). University rankings have thus become powerful technologies of the metricization and economization of today's academia.

In terms of metricization and economization, university rankings are similar to the JIF: They create varied metrics that are used to shrink understanding of the results of research practice mostly to publications and to support various financial decisions. As a form of global evaluative power, university rankings also play a substantial role at the national and local levels as reference points for laying out the aim of developing the higher education and science landscapes.

4.2 National Research Regimes and Incentive Systems

Turning now to national level, I show that states are ones of key actors in transforming academia's aims and tools through metrics and economization. The existence of evaluation regimes can be observed in all continents and regions, from the Global North and Global South to the Global East, each with its own specificity. For instance, in Europe, one notes the popularity of research evaluation systems as part of performance-based research funding systems. In this part of the chapter, I describe two types of research evaluation systems. First, those which are combined with national performance-based research funding and established by governments like those of Australia, Denmark, Finland, Norway, Poland, and the UK. Second, I discuss research evaluation systems that affect the overall national higher education and science sectors in other ways, for instance, by providing monetary rewards systems and national versions of the WoS database – as in China and Russia.

National evaluative powers are designed to assess not only institutions but also, as in Italy, Spain and Poland, to control and assess individual researchers through national systems of academic promotion (tenure). Focusing on the case of Poland, I show how national, centralized systems of academic promotion are metricized and economized. In addition, monetary rewards systems that operate at both the national and institutional levels, for instance those in China, Mexico, South Africa, and Turkey, can also be understood as forms of research evaluative power. In these instances, a system of monetary incentives as a tool of economization is combined directly with bibliometric indicators to improve the efficiency and productivity of individual researchers with the aim of improving a university's position in various rankings. In this way, incentives become technologies of evaluative power.

4.2.1 The Research Excellence Framework in the United Kingdom

Numerous studies show that the REF has an impact on labor markets, research autonomy, and diversity and that it affects organizational units and fosters forms of isomorphism that result in more homogeneous research fields and a focus on research impacts (Bence & Oppenheim, 2005; Marcella et al., 2016; Pardo Guerra, 2020; Wieczorek & Schubert, 2020). Although it is part of the British performance-based research funding system, the REF has inspired many other national systems and is still used as yardstick against which to compare them. However, the REF remains a unique approach that is based on peer review, not on metrics. Thus in terms of its metricization, this system ranks lowest among those discussed in this chapter. Yet the degree of economization within it is quite significant because the use of the REF determines funding distribution. While various metrics are used, for example, to limit the number of publications submitted by an institution, evaluation itself is not mediated – at least officially – by bibliometric indicators or international citation indexes. When the REF was first implemented, there were numerous discussions about whether it should be carried out solely with the use of bibliometrics (or using bibliometrics to inform peer review), in order to reduce its cost and ensure consistent end results (Bornmann & Leydesdorff, 2014; Oppenheim, 2008). The British system spurred the rise of numerous national research evaluation regimes, although it is still perceived as a unique system based on peers as opposed to other systems such as those in Denmark, Finland, Norway and Poland, in which bibliometrics play a decisive role. The REF has also inspired several policy instruments, for instance in Australia (Donovan, 2008) and Hong Kong (French et al., 2001), and has put assessment of research's social impact on the global agenda.

In the UK, research assessment exercises were introduced against the backdrop of a great expansion of the higher education system and of access to it. Before the introduction of the first exercises, institutional funding for research was allocated largely on the basis of the number of students, and it was assumed that all academics conducted research. The first forerunner of the REF, the RSE, was introduced in 1986 in the context of the UK's poor economic performance during the preceding years. The government that was elected in 1979 made significant cuts to public expenditure, including to universities. In 1982, the University Grants Committee (created after the end of World War I to provide a mechanism for channeling funds to universities) held an exercise the results of which were used to distribute funding and protect those universities with the highest scores. As a consequence, some universities were subject to as much as a 30% reduction in their budgets (Arnold et al., 2018). In 1986, the University Grants Committee, chaired by Peter Swinnerton-Dyer, launched the RES in response to criticism for a lack of transparency during the 1982 exercise. On the basis of its results, about 40% of institutional research funding was distributed.

The first exercise in 1986 included only traditional universities. After the Further and Higher Education Act was introduced by the Thatcher government in 1992, "new universities" (polytechnics) also became eligible for inclusion in future exercises. This Act was a clear manifestation of the evaluative power of the state. Through it, "old" universities were disciplined and subjected to new evaluation regimes while "new universities" joined the competition for the same pool of resources. This strategy was successful as all universities, both old and new, began to implement strategies to become more research and publication oriented. Pace (2018) suggests that the roots of the RAE might also be connected with Oxford University's refusal in 1985 to award Thatcher an honorary doctorate.

RSE aimed to increase research effectiveness and to allocate funding to the most productive universities. Moreover, in line with the logic of New Public Management, universities had to introduce managerial practices including project-based employment, benchmarking, and output orientation (Wieczorek & Schubert, 2020). The UK research evaluation system has evolved over the course of many cycles. Subsequent exercises were conducted in 1989, 1992, 1996, 2002, 2008, 2014, and 2021. Each cycle has resulted in both major and minor changes. These include the introduction in 1989 of a peer-review procedure mediated by subject-specific expert panels, changes to the number of units of assessment by which universities were evaluated, as well as to the number of publications included in a submission for research-active staff or assessment criteria. By 2002, university departments were ranked and a grading system from 1 to 5* (with some changes during each exercise) was introduced that was based on national and international criteria. Since 2008, results are provided in the form of a "quality profile" and individual outputs (e.g., journal articles) are given one of five possible rankings: from unclassified through "one star" up to "four stars" (internationally excellent).

Even though neither bibliometric indicators nor international citation indexes are officially used in the REF, the metricization of research is highly visible in another way. In the REF, only a part of the publications produced by academic staff in a given institution is evaluated. Thus, in the department's submission to the REF, only the best quality publications are submitted. This means that only researchers who have publications of sufficient quality ("three" or "four star") are selected to submit. In this way, the quality of outputs and therefore the quality of researchers gets translated into how "REF-able" they are, which feature is treated as a benchmark of academic success (McCulloch, 2017). The tragic case of Stefan Grimm described in Chapter 2 is testament to the degree to which research has been economized. The REF is not only an instrument of funding distribution but is also used as a tool to compel researchers to bring external funding into universities and to pressure managers into redesigning internal assessment and incentives systems.

4.2.2 Evaluation of Research Quality in Poland

Poland has a large science and higher education system. Having passed through a rapid process of massification, the country has today reached the stage of universal access to higher education and is composed of almost 400 higher education institutions in both the public and private sectors. Importantly, like other post-socialist Central and Eastern European countries, part of its Soviet legacy is the separate research system it maintains that is comprised of the Polish Academy of Sciences and its numerous institutes. A distinctive feature of the Polish system is the collegial nature of its national governance structure in which the academic community exercises a substantial level of autonomy both at the national and institutional levels.

The Polish national research evaluation system is one of the longest running in the world (Kulczycki et al., 2017). It is also one of the most metricized (dozens of metrics and parameters attempt to capture every aspect of academic activity and type of research output) and economized systems (whose results have a great impact on university funding). After the fall of the Iron Curtain, as part of broader large-scale reforms, the Polish version of the socialist model was replaced by a neoliberal system. A forerunner to the current system was established in 1990, when a preliminary assessment of research institutions was conducted. A year later, the State Committee for Scientific Research, which combined the role of a typical ministry of science and higher education with that of a funding agency, introduced a framework for peer-review evaluation of all Polish research institutions. It was the first version of a Polish performance-based research funding system in which institutions were categorized in terms of their scientific performance. This form of quantification and metricization, which is based on the assignment of categories and corresponding labels (A+, A, B, and C) has been in operation for the past three decades. Since the first categorizations, the Polish system has undergone various cycles. The argument advanced for changing the first version was that, because it assigned too many departments to the highest category, the system devalued assigned scientific categories. Thus, the second version of the Polish system was based almost entirely on metrics which, according to policy makers, made the evaluation more objective and independent of peers.

In 1999, this second version introduced a metric-based evaluation that assigned a specific number of points to various outputs like scholarly monographs, journal articles, or grants. With this, the Polish system shifted from peer-review evaluation to metric-based evaluation in which the role of expert opinions was substantially reduced. Moreover, at this time, the State Committee for Scientific Research started compiling a national ranking system for scholarly journals (the Polish Journal Ranking) in order to support the research evaluation system. Using this second version of the Polish performance-based research funding system, four cycles of evaluation were conducted: in 1999, 2003, 2006, and 2010. Each cycle changed and improved upon the previous cycle, and the range of data, the number of categories, and definitions of parameters became more precise. In 2005, the State Committee for Scientific Research was merged into the Ministry of Science and Higher Education, where it continued its work on the Polish system. In 2010, the Committee for Evaluation of Scientific Units was established as an advisory board to the Minister of Science and Higher Education; it has since become the board responsible for national evaluation in Poland. The implementation of the third version of the Polish system, which informed the 2013 and 2017 evaluations, was made possible thanks to this reorganization. Since 1999, that is, since the first edition of the Polish Journal Rankings, the JIF has played a significant role in the Polish system. Today, data from, among others, WoS and Scopus are used to prepare instruments for the evaluation, benchmarking, and ranking of evaluated institutions. In 2019, the fourth version of the system was introduced as part of large-scale reforms in the higher education and science sectors (Antonowicz et al., 2020). In this way, Poland implemented a social impact assessment similar to the Australian and UK versions, while its list of publishers was inspired by the Norwegian Model.

During the 2013 and 2017 evaluation exercises, almost 1,000 units (i.e. faculties, basic and applied research institutes, and institutes of the Polish Academy of Sciences) were evaluated. Each institution was evaluated in accordance with four criteria: scientific and creative achievements (e.g., monographs, journal article, patents), scientific potential (e.g., the status of a National Research Institute), the material effects of scientific activity (e.g., external funding), and the activity's other products. For each evaluated item, a given institution – that is, faculty or institute – obtained a specified number of points. In this way, research output and productivity were translated into the key metric which was called "Polish points." These points serve to build institutions' rankings and to assign scientific categories. An institution's quality was expressed via one of four categories: A+, A, B, or C. The best scientific units, those in the A+ category, received much greater financing per four-year period, 150% of that allocated to those in the A category. Scientific units within the B category receive 70% of the funds allocated to A category units, while those with a C get only 40% of the funds allocated to A category, and this only for half a year. Evaluation's financial repercussions are related to the block funds that are distributed annually to scientific units, although the evaluation exercise is, however, conducted every 4 years. Therefore, the categories that are assigned become binding until the next exercise. The amount of block funds provided by the government to institutions has depended, among other things, on the scientific categories assigned, although not on the total number of points obtained, as in the Czech Republic (Good et al., 2015). A consequence of this is that the fact of being ascribed the highest scientific category can become a factor that drives institutions to implement evaluation systems based on national regulations.

Through its revision in 2019, the Polish system has strengthened its evaluative power which is highly metricized and economized and which plays an important role at both the national and local levels. Evaluation results influence not only block funding to institutions but also define which institutions can award PhDs and habilitations (tenure-like degrees). Moreover, national lists of journals and publishers, which were launched two decades ago as tools for assessing institutions, also play a role in evaluation. They are to an extent used officially in promotion procedures. In addition, it is common practice among higher education institutions to employ them in assessments of individual Polish academics from across all fields of science (Kulczycki et al., 2021b). Universities have also started rolling out monetary reward systems for their employees that are based on the Polish points system, something that was previously unheard of. These technologies of power have been internalized by Polish researchers to such a degree (Kulczycki, 2019) that they use "Polish points" to characterize the value of their work in their researcher portfolios, despite the fact that this is not a requirement. They use them for reputational or marketing purposes (Dahler-Larsen, 2022; Fochler & De Rijcke, 2017).

4.2.3 Excellence in Research for Australia

In the book's Introduction, I noted that the national research evaluation systems in Australia and Poland use similar technologies of power (e.g., lists of journal lists), although the contexts in which the systems are implemented diverge substantially. Even though the Australian system (Haddow, 2022) might look similar to the Polish one, they differ considerably with respect to the degree of metricization and economization: The Australian system is less metricized and economized than the Polish one, resembling the British REF in various aspects. In 2000, Australia foreshadowed the establishment of an assessment system similar to the RAE; its main rational being to assess the quality of research supported by public funds (O'Connell et al., 2020a). Before this, from the 1970s to the early 1990s, Australian universities had been encouraged to monitor their own performance, focusing on both effectiveness and public accountability. During the 1980s, Australia's higher education and science market was transformed by forces emboldened by the global rise of New Public Management, which resulted in a strengthening of accountability requirements (Marginson, 1997). As Butler (2003a) shows, in the 1990s, Australian researchers' publication practices along with citations of their publications were altered substantially thanks to a change in the numbers of journals indexed by the Institute for Scientific Information (which one can today refer to as "WoS indexed journals"). The key factor that must be taken into account, according to Butler, when explaining this effect, is the impact of research evaluation on the scientific community in Australia. Since 1993, all Australia universities have been required to report their publication activity and this information has been used, to some extent, to distribute a part of universities' operation grants. Although Butler's results have since been put in question (Butler, 2017; Van den Besselaar et al., 2017), nonetheless, this Australian case from two decades ago clearly shows that research evaluation regimes can have an impact on how and why researchers publish.

In 2001, the Federal Government announced a five-year innovation plan called "Backing Australia's Ability." This was a strategy to build a knowledge-based economy and to enhance the government's ability to manage the higher education and science sectors in the global context. From 2005 to 2007, the Research Quality Framework (based on peer review and panels of experts) was designed and developed to be replaced in 2008–2009 by Excellence in Research for Australia (ERA). The latter implemented new technologies of power such as journal rankings (like in Poland), which divided them into four categories (A*, A, B, and C) and used these for assessing universities' performance (Vanclay, 2011). The roll out of the ERA was intended to bring about a full-scale transformation of Australian universities, planting an audit culture at their core (Cooper & Poletti, 2011).

Three subsequent rounds of the ERA were held, in 2012, 2015, and 2018. At the time of its implementation, British universities had already experienced more than twenty years of national evaluation. Although Australian universities had been reporting publication activity, the ERA still constituted a major change whose purported aim was to replace the old, cumbersome, and opaque Research Quality Framework (Cooper & Poletti, 2011) that was actually never rolled out across Australia (there were just trials at a few universities). As a result of this transformation, the mechanisms of evaluation and performance-based research funding came to be predicated almost exclusively upon quantitative indicators.

One can qualify the Australian system as moderately metricized and economized because evaluation within the framework of the ERA compares the performance of Australian universities against national and international benchmarks, in terms of the distinct disciplines (as in Poland after 2019). The results of evaluation are compiled by expert review committees who inform themselves by reference to a range of indicators. Thus for many scholars, the ERA is a metric-based assessment (Woelert & McKenzie, 2018) in that, for example, it focuses on citation profiles and the quantity of highly ranked journal articles. While ERA results do not directly influence the allocation of funding for research, they are influential in shaping institutional strategies for performance assessment and internal promotion (O'Connell et al., 2020b).

Social impact assessment, a separate initiative called "Engagement and Impact Assessment," is an interesting example of how evaluation and implementation of impact assessment is a back-and-forth process. Williams and Grant (2018) analyze how the UK and Australia designed and implemented their social impact assessment regulations within their national research evaluation systems. In studies on research evaluation, it is widely accepted that the UK introduced the first system for social impact assessment. And yet, as Williams and Grant show, Australia had in fact been on course for such an exercise several years earlier. However, due to political changes, the impact assessment was suspended, being reactivated a few years later. Thus that which had originally been designed in Australia was finally implemented there too, although in a process mediated by the British context. However, these discussions of the process in the Anglo-Saxon world do not even countenance the possibility that assessment of social impact has a much older (i.e., Soviet) history as well as distinct trajectories in the present

4.2.4 *The Norwegian Model and Its Applications*

In 2002, Norway implemented a performance-based funding system for the higher education sector. The system is indicator based and was extended with a publication indicator three years later, which has often been called the Norwegian model (Sivertsen, 2018a). Although its main function is to allocate institutional funding and Norway has another national system directly aimed at providing policy advice based on peer-reviewed research evaluation, it serves well as an example of a system designed for national purposes which has also influenced or inspired the roll out of institutional funding-schemes in neighboring countries. Before the Norwegian Model was implemented, funding of higher education institutions was based mostly on indicators representing research, that is the number of doctoral degrees, amounts of external funding, and the indicator depended on the number of tenured staff. Following a general trend in Europe, the government then initiated a system that would represent research activity more directly (Sivertsen, 2018b). The purpose of this model is to redistribute basic research funding among institutions in accordance with scores on a bibliometric indicator that counts scholarly publications. However, although the indicator is used to distribute some funding, the way in which it has been designed and is used, as well as the modest share of the budget it is attached to, means that the metricization of the Norwegian system is relatively low, and the same holds for its economization.

The Norwegian Model is based on three pillars. These consist of a comprehensive national database of bibliographical records (not limited to WoS and Scopus publications), a publication indicator with a system of weights combined with lists of journals and publishers prepared by expert panels, and a performance-based funding model that allocates a small portion of annual direct institutional funding according to "publication points" – where one publication point represents around 3,000 Euro (Sivertsen, 2018b). It is distinctive from most other metric systems by allowing for bottom-up influence of academic communities in its design and maintenance, by ensuring that all peer-reviewed scholarly publications (in journals, series, and books) in all languages are included according to a definition agreed upon by all fields of research, and by balancing between different publishing traditions when measuring productivity across all fields.

In this system, the emphasis is on a balanced measurement of research performance across different fields and the main tool designed for this end is the bibliometric indicator expressed in the number of publishing points. This indicator is not based on citations but on the number of publications and the quality of the scholarly publication channel (e.g., the journal or book publisher), which is captured in the quality level in the national list of scholarly journals and publishers. The publication channels are evaluated by expert panels to overcome various limitations inherent in bibliometric indicators which make comparison across research fields very difficult. All journals and publishers counted in the system are ranked in one of two levels and in the funding model the publication points are assigned to peer-reviewed outputs according to the quality level of the channel. For example, an article in a Level 1 journal is equal to one point and one in a Level 2 journal to three points. A monograph in a Level 1 publisher gets five points whereas one in a Level 2 publisher is awarded eight points. Level 2 constitutes at most 20% of a subject area's total scientific publications and represents the most prestigious and desirable publication channels from the Norwegian academic communities' point of view, which in fact is manifested in the higher number of points they accrue. Publication points are fractionalized according to the number of authors (Aagaard, 2015). However, from its very inception, the designers of this model have stressed that it should be used only for funding of institutions, not assessing individual researchers (UHR, 2004).

The Norwegian Model is moderately metricized and economized and could be categorized as the most balanced system among those analyzed in this chapter. While being a national policy tool, it has inspired other countries to either adopt it or to reform their own systems. Thus the Norwegian Model plays an important role not only at the national and local planes but also at the global one. Both Denmark (Aagaard, 2018) and Finland (Pölönen, 2018) have adopted the Norwegian model, modifying it to construct national versions suited to Danish and Finnish policy goals, contexts, and the needs of their higher education and research landscapes. Moreover, some of its policy instruments have inspired reforms of both the Flemish Belgian models as regards the evaluation of social science and humanities (Engels & Guns, 2018) and the Polish model's system for compiling its list of scholarly book publishers (Kulczycki & Korytkowski, 2018). Other examples of the Norwegian Model's influence abroad include its use by University College Dublin (Cleere & Ma, 2018) in Ireland and by several Swedish universities (Hammarfelt, 2018) as an institutional performance management tool. Although Sweden has not adopted the Norwegian Model at the national level, where the indicators for institutional funding are based on publications and citations in the WoS, the universities have preferred to use some of its elements for internal funding and assessment purposes in combination with other measures like the JIF.

In 2008, Denmark was the first country to use the Norwegian Model to design its own national system called the Danish Bibliometric Research Indicator (BFI). As in Norway, the motivation for establishing a country-wide performance-based funding system was to tackle the challenge of distributing core funding among Danish universities. Until that point, their system had been based on a historically conditional distribution key that did not factor in the research quality and efficiency of higher education institutions. Despite the fact that the Danish model appears almost identical with the Norwegian, a variety of local modifications have been made, and a number of problems are evident as regards implementation – for example, a lack of clear objectives on the government's part (Aagaard, 2018). The second adaptation of the Norwegian Model to the national level occurred in Finland in 2015, when a quality-weighted number of publications was introduced as an indicator to the universities' funding model. As in Norway and Denmark, the Finnish publication indicator is based on comprehensive national publication data and the list of journal and publishers prepared by expert panels. However, one finds various modifications related to the number of panels, the number of quality levels of publication channels, and the publication weights for outputs used in the funding model (Pölönen et al., 2021).

Mouritzen and Opstrup (2020) highlight the fact that Danish government started building its performance-based research funding system with the idea of measuring the quality of research, but ended up with an indicator whose purpose was to *promote* the quality of research. And yet, one cannot separate the objective of measuring quality from that of promoting it. In other words, if we take into consideration the observations that fed into Campbell's law, one understands that the measurement of results is at the same time the promotion of that which is being measured. In light of this relation, Schneider (2009) argues that, by devising distinct quality levels, the Norwegian Model encourages institutions and researchers to publish in the most prestigious publication channels. The publication of the

Norwegian Association of Higher Education Institutions, which developed the indicator in Norway, stated explicitly that "The purpose is to create a performance-based funding model for research and to encourage more research activity at universities and university colleges" (UHR, 2004, p. 3). In Finland, the purpose of introducing weights based on publication channels to the funding model was to strengthen the quality aspect of indicator so that it would not only recognize and encourage quantity of publications (Pölönen et al., 2021).

Thus, this bibliometric indicator of publishing points serves both as a metric that controls universities and as an economic incentive because, in the case of Norway, points can easily be translated into an amount of money. Such metricization and economization of research facilitate the establishment of various forms of the evaluation game. In addition, they also encourage the use of the indicator not only for distributing funding and comparing publications but also for steering and changing research practices and publication patterns. In this context, it is interesting to note that the Danish Ministry of Higher Education and Science[1] announced on 3rd of December 2021 that the ranking of journals, that is the Danish BFI, will no longer be maintained as of 2022. Also in Norway, a committee established by the Ministry to review funding model has published a report[2] in 2022 suggesting that the number of indicators would be reduced from eight to two. If this plan is adopted, the publication indicator would no longer be used as basis of allocating institutional funding in Norway. At this stage, however, it is difficult to predict how such a gap left by the publication indicator will be filled with other bibliometric information and whether it will increase the metricization of the evaluation system and use of less balanced indicators at the different levels.

4.2.5 Research Evaluation Procedures in China

In 2018, China was named the top producer of scientific articles in the world. According to the analysis by the National Science Foundation in the US that used the Scopus database, Chinese researchers published more journal articles per year than any other single nation (Tollefson, 2018). This does not necessarily mean that Chinese scholars suddenly started publishing more papers. Rather, it shows that they started to publish more and more in journals indexed in international databases which are acknowledged, from the perspective of the Global

---

[1] Ministry of Higher Education and Science, Denmark, December 3, 2021, https://ufm.dk/en/research-and-innovation/statistics-and-analyses/bibliometric-research-indicator. Online access: July 19, 2022.
[2] Finansiering av universiteter og høyskoler
Rapport til Kunnskapsdepartementet 17. mars 2022 fra et utvalg nedsatt 9. september 2021, www.regjeringen.no/contentassets/6c4c7be66d5c4a028d86686d701a3a96/f-4475-finansiering-av-universiteter-og-hoyskoler.pdf, Online access: July 19, 2022.

North, as respectable and prestigious. The pertinent question is how this change in the top position (from the United States to China) occurred, and what role the evaluative power of the state played in it.

China is a unique example of a country in which some elements of the socialist model of research evaluation are still used, and the distribution of funding for key research sectors is based on medium- and long-term strategic plans (Li, 2009). And yet at the same time, these elements are combined with the neoliberal approach to research evaluation that is evidenced in strong metricization and economization based on bibliometrics indicators. Over the decades, Chinese technological development has been driven by five-year plans inspired by Soviet approaches. The first plan was announced in 1956, and by 1965, China followed a system of central planning similar to that the USSR. During the Cultural Revolution (1966–1976), Chinese research was almost entirely frozen through the policy of shutting down universities and sending professors to the countryside to work in the fields. During the revolution, research capability was diminished and isolated from international research communication, with Chinese scholars contributing only a few publications to the market of global knowledge production (Horta & Shen, 2020). During the mid-1970s, after the end of the Cultural Revolution, the Chinese higher education and science sectors underwent profound reforms and development connected with the launch of the "four modernizations" – of agriculture, industry, science and technology, and national defense. Although it remained a communist country, China began to move toward a market-oriented research system and launched a variety of science policies in the second half of the 1990s whose objective was to realize one of Deng Xiaoping's sayings that "science is the first productive force" (Serger & Breidne, 2007).

In China, all reforms have a nationwide character and are initiated by the central government. In the mid-1990s, China launched Project 211 which aimed to raise the quality of university research and their contribution to socio-economic development by providing extra funding to a hundred universities. This project was followed by Project 985, which was designed to establish world-class universities in the twenty-first century. It was the outcome of a political decision by the government to catch up with the most developed countries in terms of science. For this purpose, the state needed metrics both to determine the current status of Chinese science and also to set goals to be achieved. Therefore, it is no coincidence that the first global ranking, that is, the Shanghai ranking, appeared in China. The intention behind it was to help determine the position of Chinese universities in relation to global and world-class universities. The Project 985 was abandoned in 2015, and the current project is called "Double first class university plan" (Yang & Leibold, 2020).

Although these projects were targeted at both the development of science and the economy, some studies argue that only the number of international publications increased, and that there was no increase in publications in Chinese scholarly outlets that could contribute to the transfer of technologies to industry (Yang & You, 2018). In 2020, China initiated a new wave of reforms of the higher education and science sectors. After becoming a global leader in scientific publications, the country is starting to focus on publishing major research results in Chinese journals (Nature, 2020). The power that large and resource rich states like China hold enables them not only to reform their national landscapes but also to shape power relations in the global scientific arena. Thus China has announced a strategy of balancing further internationalization with domestic needs, and metrics-based evaluation with peer review (Zhang & Sivertsen, 2020). For the past thirty years, the Chinese government has made great strides to shift Chinese science from the periphery to the center, whose aim was explicitly described as a way of modernizing the country. This modernization through the internationalization of science has been carried out through metrics and various economic incentives which were designed at the institutional level but became widespread across the whole country.

The key metrics relied on for locating Chinese science in the global scene are based on WoS indicators (Jin & Rousseau, 2005) eventually allowed China to surpass the United States as the largest national contributor to international scientific journals. From the 1980s onward, the international visibility of Chinese research was assessed on the basis of the number of WoS publications. This emphasis on publications in channels indexed in the WoS has been long been referred to as "SCI [Science Citation Index] worship" (Zhang & Sivertsen, 2020). This is because publications in journals indexed in the JIF and those papers with many citations in the Essential Science Indicators have become the core criteria for research evaluation, staff employment, career promotion, awards, university rankings, and funding allocation in China.

The intensive use of WoS indicators to increase internationalization and China's position in the global science and higher education landscape is widespread across the great majority of Chinese institutions. In the early 1990s, universities and research institutions initiated the cash-per-publication reward policies (Xu et al., 2021) and, as Quan et al. (2017) report, Chinese universities offer cash rewards that range from 30 to 165,000 USD for a single WoS index publication. Initially, this system applied mostly to the natural science and STEM (Science, Technology, Engineering, Mathematics) fields; with time, however, the social sciences and humanities were also included within it (Xu, 2020). The monetary reward system (until 2020), combined with China's tenure system (also based in many institutions on WoS indicators [Shu et al., 2020]), has been widely used to promote research productivity and increase the number of WoS indexed papers with Chinese affiliations. Thus Chinese universities have adopted standards from the Global North as regards the evaluation of domestic knowledge production.

The objectives of scholarly communication are clearly defined: publications in WoS indexed journals play a key role in any evaluation procedure (Xia, 2017). At the same time, both the government and private companies have established various journal evaluation systems and citation indexes like the Chinese Science Citation Database (CSCD) or the Chinese Social Sciences Citation Index (Huang et al., 2020). These report and collect information not only about international publications by Chinese scholars but also about papers published in domestic outlets. The CSCD, founded in 1989, is currently being managed by the Chinese Academy of Sciences in cooperation with Clarivate Analytics – the company behind the WoS, thanks to which fact it is now part of the WoS. Given the new goals set in 2020 and its strengthened domestic market in scholarly publications, observers around the world will take great interest in how China once again changes its system of research evaluation in order to redefine its geopolitical position.

In summary, the Chinese system is highly metricized (e.g. pressure on publications in WoS journals). It is moderately economized because while it plays an important role in the monetary incentive system, this occurs at separate institutions rather than at the national level. Yet Chinese research evaluation does not at the moment play an important role in the global discussion on research evaluation systems. However, as the world's largest producer of scientific articles, the country may well transform the way in which evaluation systems are perceived and used. Thus we should be attentive to any announcements by China that it is moving away from global metrics. For instance, the Chinese government has been trying to reduce the metricization of the research evaluation system by modifying science policy aims and evaluative technologies after 2020.

4.2.6 A New Russian Evaluation Model

In Russia, the transition from the Soviet organization of science to the current system began with Gorbachev's *perestroika* which, although it had practically no effect on science organization itself, nonetheless initiated its transformation (Radosevic, 2003). At the beginning of this process of change, the centrally planned organization of science was dropped in favor of a grant system (aimed, for the first time, at individual researchers), and the peer-review system began to be implemented. According to a study by Mirskaya (1995), during the first years following the fall of the Soviet Union, researchers experienced no dramatic alterations to the nature of their research. Still, half of all researchers changed or narrowed their research themes which fact might suggest that the previous system, based on ex ante evaluation of research topics, restricted their research autonomy. During the initial period of transition, the state limited its evaluative power; however, since the 2000s, it has regained its role as the main agent of change in the design of the higher education and science systems (Platonova & Semyonov, 2018; Smolentseva, 2019). At this time, Russia's financial recovery, spurred on by a rise in world oil prices, allowed the Russian government to once again start attending to science (Graham & Dezhina, 2008). In 2004, the government selected two universities and assigned them a special status which involved a specific model of autonomy and funding. From 2005 on, the government started to roll out programs whose objective was to fashion an elite higher education system and excellent research institutions. In parallel, given that during Soviet times, most research had been conducted at the Academy of Sciences of the Soviet Union, a reform of the organization of research was also initiated. Henceforth, it was government policy that universities would join the elite of research units, without, however, weakening the academy excessively, as the state was still in the process of reforming it (Kosyakov & Guskov, 2019).

In the Soviet Union, various institutions collected data and information about research productivity and conducted studies on the science and higher education sectors. The current situation in Russia in terms the fields of scientometrics and higher education research appears as a collapse. However, in recent years, with the renewed intensification of the metricization of science and higher education, these fields are beginning to bolster themselves and expand (Guskov et al., 2016; Smolentseva, 2019). According to Kassian and Melikhova (2019), "a bibliometric boom" began in Russia in 2012, when the government introduced various incentives to stimulate scientific development and to boost the number of publications in WoS journals. Furthermore, since 2012, bibliometric indicators have been used as a requirement for applicants to major new grant programs, with experts evaluating applications instructed to directly take into account various metrics. In 2012, the Russian government launched an annual institutional assessment of higher education bodies, and since then, it has been collecting and publishing about 150 indicators per institution (Platonova & Semyonov, 2018). In addition, the Academy of Sciences is subject to multiple institutional reviews that are based predominantly on bibliometrics (Sokolov, 2021).

Today, metricization, in close combination with the economization of the science and higher education landscape is as far reaching as it was during the Soviet era. However, it should be stressed that the nature of this metricization has changed radically. Two of the most representative technologies of evaluative power in Russia currently are the Excellence Initiative "5Top100" and the Russian Science Citation Index (RSCI; Mazov et al., 2018). The 5Top100 was initiated by the Russian government in 2012 to help selected universities enter the top 100 universities in global rankings, and it allocated special funding to them. Initially, the project included fifteen universities, but in 2015, six more were added. This initiative was inspired mostly by the success of the Chinese excellence initiatives (the above-mentioned Project 211 and Project 985). The Russian initiative aims at boosting the production of a key output counted in (unspecified) global rankings, that is the rate at which universities produce internationally recognized research (Lovakov et al., 2021). The indicators included in the 5Top100 rules directly reproduced the metrics used in the Shanghai Ranking, THE, and the QS World University Rankings. The other technology of evaluative power, the RSCI, is a special database on the WoS platform and is run by Clarivate Analytics, the Russian Academy of Sciences, and Russia's Scientific Electronic Library. It was launched in 2015. The RSCI is an effect of the "bibliometric boom" that began in 2012. Like other regional citation databases in the WoS (e.g. the CSCD or KCI-Korean Journal Database), the RSCI is not a part of the WoS Core Collection. This means that RSCI journals do not affect WoS metrics like the h-index or the JIF KCI-Korean Journal Database. Still, it used as a tool to monitor and control the science academic publishing landscape. Additionally, it is employed in various assessment procedures as a prestigious (because legitimized by an international company) assessment tool.

The metricization of Russian evaluation regimes is based on international metrics, not on peer review. Sokolov (2021) argues that the government favors the use of formal indicators because they can be calculated without assistance from the evaluated university and can be cross-checked with other sources. In this way, the objectivity of external sources is presented as a remedy for problems with peer review in countries like Russia itself, Poland, or the Czech Republic. There is, however, another factor that makes the use of metrics particularly tempting in Russia. When policy makers or bureaucrats have to take decisions about research funding or the organization of science, they can request that academics provide them with a collective opinion on the basis of which to take their decision. However, in practice, in the Russian system academics are divided (even within a single discipline) into various structural worlds (e.g. researchers within the Russian Academy of Sciences and researchers within universities); thus they represent diverse academic tribes (Becher & Trowler, 2001). In such a situation, it is almost impossible to create such a consolidated opinion. Decision makers can therefore hide behind the introduction of formal performance indicators, which grant them immunity: It is no longer their decision, but what the numbers say.

There are numerous points of convergence between the current higher education and science landscapes and discourses on them, which are often presented as neoliberal, and the Soviet approach. Smolentseva (2017) argues that post-Soviet policy documents evidence a primarily economic and instrumental understanding of the roles of education and knowledge, which is similar to that which dominated during the Soviet era. Thus science is conceived of as part of the context in which the achievement of economic indicators occurs, and science and research activities are measured through various indicators. Despite the radical shift from Soviet to neoliberal ideology, key attitudes toward these sectors, Smolentseva shows, have remained the same. Thus science is subordinated to the national economic agenda and various metrics are used to monitor and control it. In this way, metricization combined with the economization of research remains vital. The precise technologies of power have, however, been altered: the ex ante evaluation of research themes has been replaced by various ex post evaluation procedures, and the centrally planned organization of science has been substituted by market control. The most significant change, however, relates to the purpose for which metrics are used. In Soviet times, metrics served the realization of state plans, and metrics created abroad were of little importance. Naturally, the Soviet Union wanted to surpass the United States militarily, scientifically, and economically, but individual Soviet universities did not compete with individual American universities. Now, however, Russian science policy has accepted and adopted global metrics, international citation indexes, and global rankings as points of reference (but also as tools for a change). On the surface, Russia has similar mechanisms for controlling the quality of science and its organization to those used in other Western European countries or the United States. However, taking into account the analysis presented in Chapter 3, one understands that current Russian science policy is based on a tradition of science management and higher education that dates back several hundred years and is substantially different to that which took shape in the Anglo-Saxon world. This tradition likely has a great influence on how new metrics and technologies of evaluation power are implemented, used, and perceived by scientists. In a paper that examines the modernization of Russian science and evaluation regimes, Mikhail Sokolov (2021) expresses this distinction succinctly: "the policies within the academic sphere in Russia were never really neoliberal in the sense of relying on market mechanisms" (p. 8). It makes it relatively easy for Russia to turn its back on global trends and return to mechanisms that have survived under the surface covered by neoliberal solutions.

When I prepared the final version of this book for publication, the world of Russian science changed irreversibly after Russia's war against Ukraine in February 2022. The long-term consequences cannot yet be fully predicted, but one can see the first radical steps coming in the first half of 2022. These steps are cutting Russia off from global indicators and returning to domestic metrics. The Russian Deputy Prime Minister Dmitry Chernyshenko on March 7, 2022 announced that it plans to cancel existing requirements for scholars to be published in WoS or Scopus-indexed journals (Else, 2022). What is also interesting in this situation is that Chernyshenko asked the Russian Ministry and Higher Education to introduce its own system for evaluating research. Valery Falkov, the Minister of Science and Higher Education of Russia, responded to this call and in the official message published in the ministerial website.[3] It is highlighted the need to reduce the weight of bibliometric and scientometric indicators in research evaluation and to focus on national interests in the field of scientometrics and publishing. One of the proposed ideas is to create a new national list of journals and conferences and increase the weight of Russian publications.

Global evaluative power actors were quick to react to this situation. Clarivate, announced on March 11, 2022,[4] that it would cease all commercial activity in Russia and suspend the valuation of any new journals from Russia and Belarus. A week earlier, on March 4, 2022, Paul Howarth, chief executive of *Times Higher Education*, announced that in the next edition of the THE ranking THE, Russian universities will be given "less prominence." In the same announcement, however, producers of THE ranking revealed their firm belief in the objectivity of the rankings, thus hiding the fact that rankings describe the reality of the higher education sector to a small degree, and more significantly co-create and shape it: "Our rankings are based on data, and as such offer an independent view of the word as it is, both the good and the bad" (Howarth, 2022). Such a statement could only be defended by the profound and subconscious naivete of those who proclaim such views.

Turning away from global technologies of evaluation power, focusing on national solutions, and emphasizing national publications in local language can be considered a turning of the wheel in the case of Russia. However, an important question is whether this turn away from global powers will be accompanied by a return to a national system based on the *ex ante* evaluation.

4.3 Evaluative Powers Focused on Individual Researchers

As the examples discussed in this chapter show, evaluative powers at the three intersecting planes are most often designed for institutions or in order to provide bibliometric indicators and data for the higher education and science sectors. Nonetheless, eventually, these systems end up affecting individual researchers who are constitutive parts of institutions. Thus, when investigating global or national systems designed for institutions, one needs to take into account the fact that these evaluative powers will have significant consequences for individuals and will compel both researchers and academic staff managers to play the evaluation game. In this section, I introduce two different technologies of evaluative power that operate at the national level and directly impact individual researchers rather

---

[3] The Russian Ministry and Higher Education, www.interfax.ru/russia/847029, March 11, 2022. Online access: June 20, 2022.

[4] Clarivate, https://clarivate.com/news/clarivate-to-cease-all-commercial-activity-in-russia/, online access: June 20, 2022.

than institutions. It should be stressed that these two examples do not exemplify the full spectrum of evaluative powers designed at the national level and focused on individual researchers. There are other examples such as the "sexenio" in Spain which serves to evaluate the research activity of Spanish academic staff every six years (Giménez Toledo, 2016; Marini, 2018). However, the two technologies that I discuss below are representative of this type of evaluative power focused on individual researchers, and this discussion provides context for the investigation of the evaluation game that follows in Chapter 5.

4.3.1 The Centralized Academic Promotion System in Poland

Although academic recruitment and promotion systems for research staff vary across countries and regions, bibliometrics are increasingly used as a screening tool in academic recruitment and promotions (Reymert, 2020). On the one side, there is the tenure system in which "tenure" is an indefinite academic appointment whose objective is to defend academic freedom (appointment can be terminated only under extraordinary circumstances), providing stable and favorable labor conditions. Such systems run, for instance, in the United States, Canada, and the United Kingdom. On the other side, there are the systems of many European countries that are based on qualifications procedures that refer to scientific degrees higher than a Ph.D. (most often "habilitation"), which confer tenure in an almost automatic way or which are prerequisites for obtaining tenure or a professorship. For instance, in Germany, habilitation is a type of second Ph.D. that may be presented in the form of a thesis or a series of publications (Enders, 2001). In France, habilitation is one of the ways in which one can become a full professor (Musselin, 2014), while in Italy, the Italian National Scientific Qualification is a prerequisite for applying for tenure of full professor positions (Marzolla, 2016). Habilitation procedures can be regulated at the institutional level or at the national level as they are, for example, in Austria, Italy, and Poland.

Until 1951, habilitation was regulated at the institutional level in Poland (it was an authorization to teach at a given university). Since then, it has been regulated at the national level. 1951 was the year in which the Congress of Polish Science in Warsaw (mentioned in Chapter 1) was organized and reform of the Polish science sector initiated, in order to align it with the Soviet blueprint. In Poland, the habilitation degree gives researchers the right to act as supervisors of Ph.D. students. It can, furthermore, be perceived as a tenure position.

Habilitation procedures in the country were subject to substantial redesign in 2011. Since then, Poland's evaluation regime has become highly metricized (criteria are expressed by bibliometric and scientometrics indicators without a setting of thresholds to be met) and economized (an increase in a candidate's research productivity as opposed to quality has become an important assessment criterion). In 2018, the government slightly modified the procedure. Here, however, I refer to regulations published in 2011 which also apply to ongoing proceedings until the end of 2022.

In 2011, the Polish government ruled that a part of the documentation submitted for habilitation (self-promotion documents, candidates' CVs, and reviewers' reports) have to be published in order that the process can be monitored and controlled. Candidates have to submit an application in which their achievements are presented. Each procedure is conducted by a faculty or research institute and is then verified by the Degrees and Titles Committee (and since 2018, by the Research Excellence Council), which is the central institution responsible for ensuring the quality of academic promotions in Poland. When a candidate submits their application, the committee establishes a habilitation commission which issues the recommendation (positive or negative) on the basis of the reviews. The final decision is made by the faculty or research institute's scientific council. A habilitation procedure can be initiated by a researcher who holds Ph.D. and whose achievements post-Ph.D. constitute a significant contribution to the advancement of a given scientific or artistic discipline. In addition, in a clear hallmark of the economization of this procedure as regards the fetishization of productivity growth, the candidate must show that their scientific activity is substantial and was "produced" after they obtained their Ph.D.

The assessment criteria to be followed by reviewers were issued via a ministerial regulation and are divided into three categories.

In the first category, criteria are related to the field. Candidates from the arts and humanities are required to list those of their publications which are indexed in the WoS or the European Reference Index for the Humanities (ERIH). Candidates in Social Studies must list their publications included in the Journal Citation Reports (JCR) or the ERIH. Candidates from other fields can list only publications from the JCR. Moreover, a candidate can list their patents, consumer or industrial patterns and, in the arts, their artworks. In the second category, the same criteria apply to all fields. The candidate has to list: other publications that do not meet the criteria of the first category, expertise, projects, grants, invited talks, rewards, and the values of three bibliometric indicators (i.e., citations, the h-index, and Total Impact Factor) according to the WoS. The Total Impact Factor is not defined in official regulations nor is there any official formula or interpretation as to how to calculate its value. However, common practice is to use publications' 2 year JIF so that the indicator resembles "total impact" (Beck & Gáspár, 1991) or "author impact factor" (Pan & Fortunato, 2014). Among other European countries, it is only in Italy that promotion procedures are based explicitly on bibliometric indicators. The Polish approach to bibliometric indicators differs from the Italian, however, in which bibliometric indicators are applied only to the so-called hard sciences, whereas "non-bibliometric indicators" (e.g., normalized number of authored books) are used for assessing the social sciences and humanities (Marzolla, 2016). In the last category of assessment criteria, the same criteria also apply to the fields by which reviewers assess the candidate's teaching activity, international cooperation, and popularization of science. Candidates have to list: presentations given at national and international conferences, rewards, memberships of societies and editorial boards, participation in research consortia and networks, internships, supervision of students, and doctoral candidates. All information necessary for assessing a candidate according to these criteria, including bibliometric indicators, has to be presented in a document named "List of all publications and achievements" which is not publicly available and accessible only to the habilitation committee.

In a previous study (Kulczycki, 2019), I investigated the self-promotion documents presented by candidates in 3,695 habilitation procedures and found that researchers in all fields (even in the arts and humanities) used bibliometric indicators to describe the value of their research despite the fact that this was not required by regulations. Candidates relied heavily on both the JIF and the names of international citation indexes to highlight the global importance of their achievements. Both the JIF and Polish points were the two most frequently used metrics, although they were not explicitly included in the assessment criteria. With respect to the latter criteria, this practice provides clear evidence of the impact made by the research evaluation system (designed for institutions) on the self-presentation practices of individual researchers.

The evaluation power of these national procedures manifests itself in standardization and comparability (at least in theory) of scientific degrees awarded: All candidates are assessed according to the same sets of explicitly provided criteria and key documents are made publicly available. This openness provides a high level of transparency and objectivity. Moreover, the metricization of this regime provides easy to use tools for assessment, although a great deal of evaluative power still remains with peers. While this system may appear to be a desirable one, the Polish case shows that it also risks jeopardizing the balanced development of different scientific disciplines and regions in a country. Let us look more closely at this argument. The central national body responsible for monitoring and conducting habilitation procedures, the Degrees and Titles Committee, consists of full professors elected by the academic community. Members of this committee hold considerable power which they use not only to monitor and assess researchers' outputs but also to control the development of particular scientific disciplines. This they do by appointing themselves (which is not prohibited) as reviewers in academic promotion procedures. Koza et al. (in press) investigated over 12,000 habilitation procedures from the 2010 to 2020 period and identified a privileged group of researchers who control a vast majority of academic promotions by simple virtue of the fact that they sit on the central committee which, following state regulations, supervises all academic promotions in Poland. Koza et al. (in press) note that the immediate motivation for their research was provided by the information that a prominent Polish academic had completed no less than 650 official reports (plus dozens of anonymized reviews) for academic degree proceedings in Poland. Further, each reviewer receives a fee for their services, expressed as a fraction of the minimum basic salary of a full professor. This means that while habilitation procedures allow candidates to get stable academic positions, they also allow some prominent academics both to acquire considerable extra income and power within their discipline. Thus the evaluation of individual researchers can, at different levels, become a means of monitoring and controlling the development of institutions, regions, and disciplines.

4.3.2 Monetary Reward Systems

In the discussion of Chinese evaluation regimes, mention was made of monetary reward and incentives systems for individual researchers. However, in academia, this phenomenon is global and can be found in various, sometimes very different, higher education and science systems. The resurgence of interest in performance-related pay is linked to New Public Management (Perry & Engbers, 2009). However, these tools were also widely used in the Soviet Union where paying for performance was the basis for the whole "model (shock) worker" movement in which financial prizes, badges, and certificates were awarded to workers who exhibited exemplary performance.

Monetary reward systems are set up as performance incentives by which researchers receive a direct monetary reward that is proportional to their level of performance. In the majority of cases, such systems are based on task-based incentives which involve the awarding of a fixed sum for completing a task (Hur & Nordgren, 2016). In the case of research evaluation regimes, the most popular approach is to pay for the publishing of journal articles in top-tier journals, most often defined through their JIF (Kulczycki et al., 2022). The main idea underlying the logic of task-based incentives is that the linking of rewards to the completion of tasks increases researcher motivation to perform well on the task (i.e., publishing a paper in a top-tier journal). Designers of such incentives systems further assume that people as well as institutions are responsive to incentives, and that economic incentives work best.

Numerous countries around the world have implemented cash rewards (task-based incentives) for publishing in top-tier journals which are defined mostly through bibliometric indicators, as well as for winners of major research grants

(Andersen & Pallesen, 2008; Sandoval-Romero & Larivière, 2019). The use of publication records for rewarding researcher performance has a longer trajectory in North America than in Asian or European universities. In the latter cases, promotion and tenure procedures have not depended to a meaningful extent on publication records (Kim & Bak, 2016). Today, with the implementation of incentives systems, this situation is changing. The primary goal of such incentive systems is to increase the number and quality of publications by researchers from a particular country. However, studies on the operation of these systems show that their ultimate effects can be quite different. For example, Chinese universities have achieved a significant increase in the number of publications, but the increase in quality (measured by bibliometric indicators) is not as notable (Chen, 2019; Quan et al., 2017). A different situation was observed by Nazarovets (2020) in Ukraine, where publishing in Scopus-indexed journals was encouraged. There, from 2015 to 2019, more than three quarters of publications were in local Ukrainian journals. Similarly in Turkey, it appears that the support program based on monetary incentives has had a negligible impact in terms of increasing the number of papers in the WoS (Tonta, 2018). In South Africa, within a decade, monetary incentives had doubled the number of publications in Scopus but had also propelled researchers toward questionable, predatory journals (Hedding, 2019). This is not only the case in South Africa; across the globe, monetary reward systems strongly promote different forms of evaluation game, as Chapter 6 will show. Reward systems create favorable conditions for playing various forms of evaluation game. For instance, in China, a university professor published 279 articles within five years in a single journal and collected more than half of the total university cash rewards from his university (Chen, 2019). This is not an isolated case, as such incentive systems reinforce the Matthew effect of accumulated advantage (Merton, 1968): The rich get richer and those who have good (in bibliometric terms) track records get more rewards and grants which allow them to produce more papers, thereby generating further incentives. This is why the amassing of rewards and the acquisition, through gaming, of ever more financial bonuses is so compelling.

The observed effects of such systems are not only negative. For instance, in South Korea, both the number of publications and the average impact factor of journals increased through financial rewards based on performance incentives (Bak & Kim, 2019; Kim & Bak, 2016). The Danish monetary reward system also can be characterized as an effective science policy tool because financial incentives can improve researchers' productivity (Andersen & Pallesen, 2008). Indeed, the cases of both Denmark and South Korea show that financial incentive systems can work when they are not in conflict with the professional standards of researchers. That is to say that first they do not encourage publication in journals that are not attractive from the perspective of disciplinary loyalty, as described in Chapter 1, and second, that researchers feel that these systems serve to support and not control them.

Perry and Engbers (2009) argue that performance-related pay frequently fails to produce the expected medium-term changes in employee perceptions that are necessary for shaping their motivations. They further point to a variety of contextual factors that have impact on the effectiveness of incentives; especially important among them is the type of public service industry involved. There are thus fundamental differences in the ways in which different groups, for instance (1) workers (e.g., factory workers or miners) versus (2) public sector employees (e.g., social workers or researchers) can be motivated to work harder. The latter group is driven by additional factors (other than financial) that motivate them to increase the productivity of their work. Such factors might be a sense of performing an important community function, helping others, or simply research curiosity. Thus, performance-related pay or task-based incentives may have a greater effect at lower organization levels, where job responsibilities are less ambiguous. Moreover, researchers are usually perceived as self-motivated and autonomous which in fact, at first glance, might mean that economic incentives do not appear to be the most suitable means for changing their research behaviors. And yet the logic of systems adopted in various countries shows that designers of evaluation regimes assume that researchers are no different from people in other areas. In planning to implement monetary reward systems, decision-makers should take into account the fact that such systems may not only alter employee motivation but also how they prioritize their tasks in and out of work. Hur and Nordgren (2016) document how being exposed to performance incentives changes the way in which one values money and other incentivized rewards. This can have significant repercussions in daily life because, as these authors highlight, how much one values a reward affects how one uses that reward. For example, Hur et al. (2018) show that when employees are paid for performance, they prioritize spending time socializing with work colleagues at the expense of spending time with friends and family.

Monetary reward systems are clear manifestations of the integration of metricization and economization. Journal articles are treated as key products made by researchers; thus journal metrics are used to indicate researcher goals. According to their promoters, the completion of these goals generates monetary rewards and allows institutions to gain better positions in diverse rankings and evaluation regimes. However, the situation is by no means as positive as such statements suggest. For example, although task-based cash incentives in China, South Korea, and Turkey did increase submissions from the three countries to the journal *Science* by 46% (Stephan, 2012), overloading reviewers in the process, they did not increase the number of papers accepted from those countries. Research evaluation regimes can easily influence research practices in scholarly communication, especially when they are designed directly for individual researchers. However, the response to incentives is not always, if it is ever that envisaged by system designers.

* * *

Across the world, research evaluation systems are integral parts of the science and higher education sectors. Global evaluative powers like the JIF or university rankings THE infiltrate national and local systems and inspire the introduction of ever more innovative monitoring and control systems. Evaluation regimes, as I have shown, vary in the extent of their metricization and economization. It follows that the effects produced by these systems and therefore also the forms that the evaluation game takes, also vary. In order to compare and highlight differences in the levels of metricization and economization in those systems analyzed in this chapter, I have created a heuristic based on two axes representing the forces of metricization and economization (Figure 4.1). Each of the systems analyzed in this chapter is to some extent metricized (e.g., in terms of criteria used or the way in which evaluation results are presented) and economized (e.g., even when there is no direct link with funding, economization can be manifested in the way in which assessment criteria are designed).

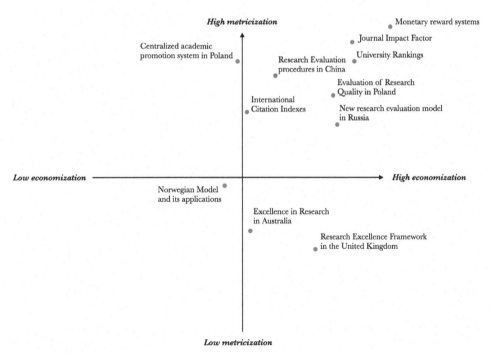

Figure 4.1 Metricization and economization of research evaluation systems.

This heuristic shows that I characterize two-thirds of global evaluation systems (JIF and University Rankings) as highly metricized and economized, much like the national evaluation systems and procedures in Poland and Russia. On the other side, only one system (the Norwegian Model and its applications) is located in the approximate center of the diagram. Two core observations might lead one to the assumption that the higher a system's level of metricization and economization, the more people will be encouraged to play the evaluation game. The first is the finding that the use of metrics to manage or monitor social practices generates a variety of unexpected reactions; and the second that researchers, just like other people, respond to a variety of economic incentives and stimuli. In Chapter 5, putting the concept of the evaluation game to work in real-world settings, I demonstrate the utility and uniqueness of this analytic tool. I do so by exploring how key actors (players), that is institutions, managers, and researchers, play various types of evaluation game, and how these are related to levels of metricization and economized in the systems in which they work.

5

Players and the Stakes

Emily published only a few papers in the period of 2013–2015. During that time, she moved from the University of Strasbourg to the University of Cambridge. These three spare facts over a three-year period amount to the entire record of her scientific life, so it is difficult to say whether or not she was a productive researcher. Emily only existed, however, as Wadim Strielkowski's false identity. Given this, one might say that she was a quite a success, being a co-author on several publications in journals indexed in key international databases. Nonetheless, in comparison to Strielkowski himself, hers was not a great achievement: As a junior lecturer at the Faculty of Social Sciences of Charles University in Prague, he had published and coedited seventeen monographs and more than sixty journal articles in just three years (Brož et al., 2017).

Many of Strielkowski's articles were published in so-called predatory journals and all (except one) of the monographs were self-published (Brož et al., 2017). However, some of the papers were also published in journals indexed in Scopus and WoS databases, thus they were counted in the Czech research evaluation system. More important here is the fact that the university that employed Strielkowski was awarded its corresponding share of funding because at that time in the Czech research evaluation system, publications were translated into points that in turn were translated directly into a sum of koruna (Good et al., 2015). Brož et al. (2017) highlight how Strielkowski received salary bonuses thanks to his impressive productivity, as based on articles counted in the research evaluation system.

Emily, his imaginary co-author who was apparently affiliated with prestigious European universities, served to give Strielkowski's publications a more serious and attractive air (Brož et al., 2017). Yet, this was not the only way in which he increased his productivity. Strielkowski acted as an editor in chief, editorial board member, and publisher of some of the journals in which he published (De Rijcke & Stöckelová, 2020). Moreover, he published some papers with his superiors, for example, the Vice Dean and Head of the Institute in which he worked. These figures in the university administration were evaluated at the same time as those managers and researchers responsible for their institutions' evaluation results. Thus, for both Strielkowski and them, publishing multiple eligible papers in their names and with institutional affiliations was a convenient way of achieving two goals simultaneously.

Strielkowski left Charles University in 2015 and founded for-profit companies that organize courses on how to get published in journals indexed in Scopus and WoS. In a publish or perish research culture, this skill may well be necessary in order to survive in academia. In some countries, however, in which evaluative power produces strong monetary incentives, publishing in such journals may serve as just another way of making a profit, as was shown in Chapter 4. Stöckelová and Vostal (2017), who brought the Strielkowski case to the attention of those outside of the Czech academic community, argue that this is a story of personal ambitions and institutional survival strategy within the framework of the systems of research funding and evaluation. Strielkowski's practices were in compliance with the Czech research evaluation system's rules – with the problematic exception of his imaginary co-author – and, as Strielkowski publicly argued in his defense, all his publications were perfectly legitimate.

Strielkowski understood how the Czech research evaluation system works and simply played the evaluation game. Everything was legitimate, his faculty received funds as a result of his publications and he also got some extra money. This hyper-efficient productivity attracted the attention of other researchers and managers. When a player finds out how to get around the system (while staying perfectly in line with the rules) and gains more funds, then other players feel cheated or gamed. This was the very moment in which players and decision makers started considering how to improve the system and to prohibit such strategies. Thus the fact that such a strategy is permissible in the system is viewed as highlighting a gap within it, rather than being a sign of the problematic nature of the system itself.

In the Strielkowski case, such strategies were based on publishing in dubious journals which were nonetheless counted by the system, as well as on self-publishing monographs. This latter strategy of self-publishing or publishing monographs with questionable publishing houses was also occasionally used by Polish scholars to increase the evaluation results of their faculties. The Ministry of Science and Higher Education in Poland noted that the record holder, that is single researcher, had published 53 (sic!) monographs in the 2013–2016 period (Ministerstwo Nauki i Szkolnictwa Wyższego, 2018). Because of this case as well as other factors, the Ministry introduced significant changes to the Polish system and, from 2017 onwards, a reduction in the number of publications that could be submitted by a single researcher. Under the new regulations, the maximum number of publications is set at four (using fractional counting). Previously, as in the Czech system, there had been no limitation. Thus, the measure became a target, and the evaluation game took place at the lowest possible cost, that is through self-publishing.

* * *

Strielkowski's case shows how diverse actors in the field of science can be harnessed to meet evaluation demands or to exploit the incentives introduced by research evaluation systems for private gain. In this story, we have a researcher who is able to work out and navigate the rules of the evaluation system so skillfully that not only does he get financial rewards but also his superiors want to join in his activities so that the institutions they manage can achieve better evaluation results. This story of the fictional scientist and Strielkowski's practice provide a good starting point for describing the key players and the stakes they play for in the evaluation game. Moreover, the story brings together two phenomena that at first glance may appear to be identical, that is, gaming and playing the evaluation game. However, as I show, these terms refer to substantially divergent practices that should be distinguished because both their causes and ethical outcomes are different.

So far in the book, by focusing on research evaluation regimes, I defined the evaluation game and described how science as a system has been transformed by metricization and economization. The argument was developed across three planes: (1) the factors of evaluative power that shape scholarly communication (metricization and economization); (2) the historical context of research evaluation system implementation (socialist and capitalist), and (3) the geopolitical context (global, national, and local). In this chapter, I foreground the utility and uniqueness of the evaluation game as an analytic tool that is sensitive to geopolitical perspective. I do so by expanding on three further lines of argument. First, I define the conditions under which it is possible to identify evaluative power as a cause of transformations in scholarly communication. Second, I demonstrate why "playing the evaluation game" should not be identified with "gaming." Third, I describe key players who play the evaluation game and consider the various stakes in the game.

5.1 The Causal Nature of Research Evaluation Systems

In this book, I frame research evaluation systems as technologies of power belonging to the state or organizations, whether global or local. Thus, I assume that the state has the power to design and introduce such systems that in turn redefine the conditions in which higher education and research institutions as well as academic staff operate. On what basis do I justify the above claim that a research evaluation system can be the cause of change in science or scholarly communication? In other words, what prompts me to attribute behavioral causality to research evaluation systems and claim that metrics transform scholarly communication?

First of all, one should not forget that all indicators or research evaluation systems produce effects in and of themselves. These effects occur when those systems are introduced and implemented, and their rules internalized by academic staff. Among the factors that allow research evaluation systems impact scientific practices are those most closely associated with transformations in scholarly communication. First and foremost, researchers are dependent on the institutions that fund them, whether those institutions are public or private. Thus, a change in regulation by the institution will force a change in the behavior of researchers. If institutions or governments introduce incentive systems, then researchers will respond to them, just like any other social group. Research institutions are also subject to constant evaluation processes. Through the publishing of rankings and detailed data on ranking results, institutions and researchers begin to compare themselves, knowing that they will be evaluated and checked in the same way by other actors in the scientific field.

The attribution of causality is a problematic that has been debated and analyzed for decades, and the use of causal claims by researchers in the social sciences is fraught with problems when claims are not supported by experimental data. In studies on research evaluation and science policy, it is often asserted that a given policy instrument has produced an effect, changing the behavior of individuals or entire social groups. It may also be alleged that particular public policies have contributed to the transformation of various social and cultural practices. One may come to believe that evidence for such causal relationships is already well established in many fields. However, if we examine these claims, we will likely find that they are based neither on actual experiments nor on strict and controlled empirical data. Instead, they rely on their author's extensive knowledge of a specific study on the historical and current context of a particular policy's implementation. Researchers in science and public policy often brush aside the need to justify their attributions of causality, which they view as belonging to the domains of philosophy or metatheory rather than social research. However, even where social experiments (such as those frequently conducted in social psychology) are inappropriate, because the object of analysis is a long-term social process, one should still remain aware of the difficulties in ascribing the state with an evaluative power that is the cause of determined processes.

Aagaard and Schneider (2017) are among the few in the field of research evaluation studies who highlight the challenges facing researchers who want to link research evaluation systems with changes in researchers' productivity and publication practices. In their view, there are three fundamental conditions that must be met in order for causal claims to be used in such studies: (1) precedence, (2) correlation, and (3) non-spuriousness. The first condition refers to the fact that a given system or its technologies must have been implemented before the observed effect. In other words, the cause needs to precede the effect. Determining when a particular evaluation system or rule change has taken place is usually straightforward.

However, determining when a given effect occurred is sometimes a difficult task, as such effects are complex processes. Moreover, as Hicks (2017) shows, there is a time lag between the moment in which a public policy tool is implemented, that in which this fact is communicated, and when it produces its first effects. Therefore, identifying the specific point in time at which a policy produced effects is not an easy task. And given that this is the case, it may happen that public policy instruments will be indicated as causes, even though they did not exert much influence on a change in social practices, because in the meantime other phenomena influencing those practices emerged. Thus, as Aagaard and Schneider (2017) highlight, institutions and individuals can change their behavior more or less from day to day.

The second condition, correlation, relates to the use of bibliometric indicators, as the dependent variable that is measured, to identify changes in researchers' publication patterns. This is a challenging task because databases are dynamic and characterized by strong growth over time in terms of indexed journals and the number of articles, references, and citations attributed to them. In Chapter 2, I described those characteristics as modalities of bibliometric indicators which are often hidden from users but which in actuality put in question the interpretation of indicator values and statistical relationships among them. Consequently, all modalities of the data analyzed should be known to the researchers investigating research evaluation systems, and they should take them into account when interpreting effects.

The third condition, non-spuriousness, is a challenge that all social science researchers face: How to isolate the effects of research evaluation systems from those caused by other forces? In principle, it is possible to do so only in social experiments, which in this case are not practically feasible because the implementation of evaluation regimes is a process that takes many years and engages institutions and researchers from entire countries. Moreover, when one is referring to changes in a system implemented at the global or national level (e.g., strengthening the metrics of the research evaluation system in a given country) and the consequences of these changes at the analogous level (e.g., changes in the publication patterns of researchers within a discipline or researchers from a particular country), then one is describing the situation at the macro level. The effects of research evaluation systems, however, occur primarily at the meso and micro levels: They are observed in the individual practices and actions of researchers and managers. However, not all researchers will confront the same effects, and some will not experience them at all. Thus, when a bird's eye view is taken, perspective is flattened out and nuances are omitted, which can lead to potentially incorrect attributions of causal influence to evaluation systems. For instance, Osuna et al. (2011) show that studies of the individual research evaluation system in Spain attributed the significant increase in the volume of scientific publications to the research evaluation system. And yet, as Osuna and colleagues show, this is instead a consequence of the increase in expenditure and the number of researchers in the Spanish system.

Given these conditions and the challenges faced by research evaluation scholars, it would be fair to say that all such analyses are constructs built on uncertain ground which is not fully defined and is beyond control. There is however an additional factor that makes it possible to undertake such analyses and to draw well-founded conclusions from them. I am referring here to expert knowledge and to the fact of being embedded within a particular system. This gives one knowledge about the context of the implementation of a public policy, historical knowledge of the development of a practice, and knowledge of the many different factors which could have influenced the observed effects. This positionality is a necessary element in a reliable analysis of the effects of research evaluation systems and claims that they are the cause of specific transformations in scholarly communication.

Detailed knowledge of a given higher education system and the implementation of policies enable one to observe and take into account two essential that are necessary for attributing causality: (1) effects specific to a given research evaluation system, even if similar effects could be caused by other, more global factors, and (2) time delays related to the internalization of new evaluation rules by researchers. With regard to the first factor, it should be emphasized that in order to indicate that an evaluation system has an impact on changing publication practices, to take an example, one must isolate those elements of the system that are very specific to it and that will allow one to identify atypical changes. This is important because the publish or perish research culture operating in academia pushes all researchers to begin publishing more and more. Therefore, one could not claim, for instance, that the revision of the research evaluation system in Poland in 2009 was the sole factor responsible for prompting researchers to start publishing more papers. This is because it was not only the evaluation system that motivated researchers to publish more. And yet, if one knew the context of the implementation of this system and its detailed provisions, one would also look at the volume of published scholarly books. Doing so would reveal the fact that in Poland, after 2009, the volume of monographs began to decrease to six author sheets, because in the evaluation system, only books with at least this volume were counted (Kulczycki, 2018). Another example would be the use by researchers of bibliometric indicators in CVs and various self-promotion documents in order to highlight the quality of their work. This practice is becoming more common in academia in many countries, with researchers most often using the JIF (Else, 2019). However, analysis of Polish researchers' self-promotion documents shows that in addition, they also use "Polish points" to value their research output. This is a "mark" through which one can recognize of the significant impact on researchers of an isolated instrument of science policy. Therefore such "marks," which are specific to national research evaluation systems and not directly connected with more global trends, allow one to identify the effects of determined policy instruments and claims that assign causality to research evaluation systems.

In order to identify the real effects of a public policy instrument and to understand when an effect may have started to transform practices, one needs detailed knowledge about the implementation of individual instruments and the overlaps between different public policies. It should be stressed that it would be an oversimplification to claim that a science policy announced in, for instance 2006, has had an impact since the moment of its implementation or from the following year on. Such changes must first and foremost be widely communicated to the academic community. Moreover, these changes must also be internalized by researchers who will, for example, strive to meet evaluation guidelines. In my work on designing and developing research evaluation systems, I have found that some minor changes, which may be cancelled after only a few months, are only internalized by researchers after a delay of several years. This causes many researchers to be guided by system rules that have been inoperative for a long time. Thus it is not the case that the moment a regulation is announced it completely overwrites the previous rules in terms of their validity and understanding of them. Therefore, expert knowledge and a firm rooting of research evaluation scholars in a particular system is necessary not only in order to isolate the causes of changes but also to determine the point from which a given change might affect scholarly practices.

5.2 Gaming and Playing the Game

Before I move on to outline the main players in the evaluation game, it is worth returning for a moment to the definition of the evaluation game in order to clearly distinguish the game from other practices that are often combined with it, notably gaming and fraud. It should come as no surprise that I clarify this distinction here, since the evaluation game is, as an analytical tool, a heuristic without rigidly drawn boundaries. This is evident if one attempts to analyze the actions of the protagonist who opened this chapter, Strielkowski. The line between fraud or misconduct and questionable practices like gaming and playing the evaluation game is difficult to determine. Yet, it is worth keeping this distinction as it shows that there are different motivations behind gaming and playing the evaluation game. Pressures in academia that make researchers want or need to publish ever more articles can lead to three categories of actions that may be legally prohibited or judged morally wrong in light of principles that are currently salient in science, and scientific ethos.

The first category is fraud which involves, for example, fabricating or falsifying scientific data and publishing a scientific article on that basis. One of the most well-known cases of this is social psychologist Diederik Stapel's fabrication of data for numerous studies (Crocker & Cooper, 2011). There is no doubt that such action is not only incompatible with the law, but that it also violates the values of the scientific ethos, as it undermines the authority of science and replaces the pursuit of truth with that of increasing personal gain, both professional and financial. It is also the case that this kind of fraud has long-term consequences, even when journals retract a paper that was based on fabricated data. Thus the retracted works of Stapel are still cited (Fernández & Vadillo, 2019) in a positive light within and beyond psychological literature.

The second category of actions is gaming which takes place when someone tries to maximize their gains, for example, by obtaining greater financial rewards through intentional "salami publishing" – dividing a large study that could be published as a single research paper into smaller papers. Alternatively, one might also speak of gaming in relation to the practice of employing academic staff on contracts with the lowest level of employment that qualifies them for being evaluated as an employee of a given institution. Thus, gaming is a strategy to maximize profits (including financial) which is fully in line with the rules, but it is often combined with finding loopholes in the legal system (e.g., through unforeseen but permitted interpretations of the rules). Thus, gaming, although perfectly legal, can be seen as unethical in that it violates scientific ethos. For example, splitting a single study into multiple articles just to get extra financial rewards for each article results in a lot of unnecessary publications in science and increases information overload.

The third category refers to the practice of playing the evaluation game. Like gaming, it is fully compliant with legal principles but is not as easy to evaluate in moral terms as is gaming. The goal of players of the evaluation game is most often to maintain the status quo (e.g., keep their jobs, meet evaluation requirements) by following the rules at the "lowest possible cost," not to maximize profits. Following the rules at the "lowest possible cost" is crucial in the distinction between "playing the evaluation game" and "gaming." Evaluation of the ethical aspects of playing the evaluation game, however, must take into account additional structural dimensions, that is, how the institution – whose rules are met through playing the game – provides resources for the work needed to meet its requirements. Indeed, if the rules of the game are changed in the course of the game, if the institution demands excellent scientific results but does not provide sufficient resources for research, or if the apparently decent working conditions are a fiction (e.g., researchers in the social sciences and humanities often do not have their own offices or are assigned by the dozen to a single room), then playing the evaluation game can be seen as a legitimate reaction to evaluative power. In such a situation, publishing in predatory journals just to fulfill the requirements of the evaluation system (if this kind of publication is allowed) would be seen as a response to unreasonable demands on the resources provided. In other words, fraud is always fraud and resources provided by institutions cannot change this. Nonetheless, determining whether an activity is gaming or playing an evaluation game is a more complex affair. This is because an activity (e.g., publishing in a predatory journal) may be considered both gaming and playing the evaluation game, depending on the context. Thus it can be thought of as gaming when it serves the purpose of maximizing profits, and as playing the evaluation game when it fulfills evaluation requirements (which implies that there is a clearly defined threshold), and when the stakes of the game are not so much additional financial bonuses as maintaining the status quo. This distinction is necessary because it shows how the actions of some researchers are motivated less by a desire for personal enrichment and are rather structural consequences of evaluative power.

Let us now look at how these three categories of actions which influence scholarly communication work in practice, focusing on the activities of Wadim Strielkowski. Undoubtedly, Strielkowski is a productive researcher who fights against predatory journals (e.g., Strielkowski, 2018a, 2018b), despite the fact that he is accused at the same time of publishing in them. In his arguments against his detractors, Strielkowski emphasizes the fact that all journals or publishers can easily be accused of being predatory or of low quality. Additionally, Strielkowski highlights the fact that his accusers publish in low-quality journals and that their actual publication practices are not consistent with the values that they promote (Strielkowski & Gryshova, 2018). According to Strielkowski, Czech researchers started to seek out new publication venues, especially those indexed in WoS or Scopus, when the Czech government introduced a research evaluation system based on points assigned to outlets. What is important in the discussion among Czech researchers on Strielkowski's publication patterns is that the implementation of the Czech research evaluation system provided an additional dimension of understanding and ethically (or even legally) assessing researchers' publication behaviors. There is no doubt that fabricating a co-author affiliated to prestigious universities is a rather awkward feature of the story and a practice that is definitely not necessary for publishing good research results. However, if one considers the consequences of publishing and self-publishing so many papers, then one needs to use both the analytic tools of gaming and playing the evaluation game in order to capture all the nuances of the practices in question. If an increase in the number of publications is motivated by economic incentives and the desire to obtain salary bonuses, then the self-publication of monographs in line with the rules should likely be named "gaming." However, Strielkowski published some papers with his superiors who were responsible for the evaluation results of their institutions, which in fact may provide evidence that they not only approved of but also supported his efforts at improving their institution's productivity profile.

The Czech research evaluation regime, as part of the funding system, compelled various researchers and managers to find ways of sustaining their institutions' funding levels within a new framework. In a redefined context such as this one, gaming might become playing the evaluation game because it is a reaction to structural changes. Even researchers who put the spotlight on Strielkowski's practices highlighted the fact that his case should be understood not as a moral failure but as a tale of personal ambitions and institutional survival strategies (Stöckelová & Vostal, 2017). Of greatest significance here is the fact that this strategy was implemented within the boundaries of the Czech system. Therefore, in order to capture the various nuances of researchers' and institutions' responses to evaluative power, in this book I use the concept of evaluation game rather than gaming although it is sometimes difficult to tell when one practice turns into another.

Gaming or fraud issues are often treated as ethical dimensions of individual conduct rather than as part of the functioning of the science system and its organization as a whole. The playing of the evaluation game is structurally motivated by the evaluative power that transforms researchers' working conditions. Gaming, by contrast, is always the result of individual ambition and most often serves personal gain, although, like the playing of the evaluation game, one of its consequences may also be the maintenance of the status quo. However, it may be simpler to gauge the proliferation of gaming and fraud than instances of playing the evaluation game. Gaming and manipulation are highly sensitive issues because they touch on law breaking (fraud) or acting at its margins (gaming). In relation to fraudsters, public interest follows cases of law breaking or of deceit through the publication of false scientific results. In the case of gaming, the public may be interested in people who have inappropriately, but legitimately received awards for academic achievement. However, in the case of playing the evaluation game, one can only speak of considerable amounts of faculty time and funds spent on attendance of low-quality conferences or for covering the costs of publishing in journals lacking in prestige. However, the assessment of the quality of conferences or the prestige of journals is a matter internal to specific scientific disciplines, and it is difficult for outsiders to judge whether a publication is indeed a waste of public money. Thus the playing of the evaluation game, although a salient dimension of academic practice, rarely attracts widespread public interest.

5.3 The Stakes for Four Key Players

The evaluation game is a top-down redefined social practice. As I argued in Chapter 1, practitioners such as researchers, managers, publishers, editors, and policy makers become players (or rather they are forced to be players) because they wish to at least maintain their current positions. Players play the evaluation game because various resources are at stake within it. The variety of resources differs not only across countries and institutions but also between players because not all stakes

Table 5.1 Primary motivations and the stakes for key players of the evaluation game

Players	Primary motivations for engaging in scholarly communication	Stakes
Researchers	Intrinsic satisfaction, loyalty to institution/discipline, amount of remuneration, discovery, and understanding	Labor security, priority of scientific discovery, recognition, rewards, resources for research
Institutions and their managers	Fulfilling the goals of the institution (teaching, research and development), maintaining and improving ranking position, beating the competition in the higher education market, hiring the best staff	High positions in rankings, good results in institutional assessments
Publishers and editors	Communicating science, increasing profits, publishing ground-breaking research results, attracting top scientists as authors, reproducing network of researchers, being recognized as a prestigious publication channel	High values of bibliometric indicators, composition of editorial boards and authors
Policy makers	Fulfilling policy goals, improving ranking positions of domestic institutions, fair and reasonable distribution of funds to institutions on the basis of research outputs, objectivizing research assessment	Increase in the number and quality of publications in international databases, increase in the number of institutions in global university rankings

are desired by all types of players, and not all players have the same starting point for playing. The game is an ongoing practice because once it is established, that is, once evaluation power redefines social practice, those working in academia are forced to play it, which is to say that they attempt to adapt and modify their practices in line with evaluation rules.

Let us take a look at four key players who are transforming scholarly communication by playing the evaluation game. I refer here to: (1) researchers, (2) institutions (e.g., universities, research institutions) and their management staff, (3) publishers and journal editors, and (4) policy makers. Table 5.1 presents the primary motivations for engaging in scholarly communication practices and the stakes that each group plays for when undertaking the evaluation game.

Above I am describing key actors across the vast fields of science and higher education which, even within a single country, are very diverse. Therefore, this characterization is simplified and merely highlights key features. In presenting it, my intention is to offer a conceptual framing for the three cases of playing the evaluation game that are presented in Chapter 6, rather than to claim to provide a complete overview. In order to distinguish gaming from the playing of the game and to capture geopolitical differences across games, it is first necessary, if only briefly, to characterize the key players in the game, their motivations for engaging in scholarly communication, and the stakes for which they play. Thus, the following discussion is focused on the following: players' roles in the scholarly communication system; the power relations in which they are entangled; the motivations behind their involvement in scholarly communication; the stakes in the evaluation game; the impact of the geopolitical location of players on how they play the game; and how the historical context of the implementation of science evaluation affects how they respond to changes in the science evaluation system.

Researchers

As the largest group of actors in the science system, researchers are also the largest group of players. They are the foundation of scholarly communication: Without their work, there would be no research results to be communicated through journals or scholarly books and used at universities to teach students. On the one hand, researchers have autonomy in choosing research topics, in selecting collaborators and venues for publishing their research results (Neave, 2012). On the other hand, however, they are entangled in various power relations that have broad repercussions. These render them heavily dependent on the institutions that fund them (above all universities or research institutions and grant agencies), on businesses that drive the academic publishing market, and on public policies that may change their duties (e.g., oblige them to intensify research at the expense of teaching). Implicated in these power relations, researchers are motivated both by wages and labor security as well as by a pressure to follow the collective views and values shared by their community – which can also provide them with intrinsic satisfaction (Blackmore & Kandiko, 2011; Igami & Nagaoka, 2014). It is a commonly held belief that researchers are mostly (or at least more than other workers) self-motivated and driven in their work by a desire to achieve the intrinsic satisfaction inherent in scientific discovery (Mouritzen & Opstrup, 2020). Moreover, these motivations play an important role in shaping researchers' reactions to evaluative power and the implementation of accountability regimes.

In addition, and rather self-evidently, researchers aspire to having a job that allows them to support themselves and their families. In addition, this work might (and according to scientific ethos, should) serve society by, among other things, improving living conditions, explaining phenomena, creating life-saving vaccines, or casting light on social phenomena. Therefore, in their communication of science, researchers are motivated in two ways: On the one hand, in a publication-oriented science system, they publish in order to maintain their jobs; on the other hand, through their scholarly publications, they strive to communicate with other specialists in their disciplines. These motivations are inextricably linked to two "loyalties" (disciplinary and institutional), about which I wrote in Chapter 1. In current accountability-driven academia, scholars are forced to balance these two types of motivation. Publishing academic papers is thus an expression not only of scientific ethos (i.e., sharing one's findings with the scientific community and general audience) but also of a common-sense approach to scientific work as simply a job (i.e., one publishes articles because the funding institution expects it). The question of whether academic publishing becomes part of the evaluation game depends precisely on whether the scales are tipped in favor of motivating factors that turn around the preservation of one's material position in a scientific institution rather than scientific ethos.

Researchers participating in scholarly communication practices may obtain recognition from other researchers and may also receive various rewards (both monetary and nonmonetary relating to, for example, priority of discovery). In addition, they can also meet the requirements of their employer, if such requirements are explicitly expressed in a set number of scientific publications in a given period of time. Thus, the three most important stakes in research as a job are recognition, rewards, and retaining one's job. Still, it should be stressed that the variety of possible stakes is vast, and in any research, teaching, or administrative practice, researchers may be playing for diverse resources. For instance, such stakes may involve getting the best graduate students and postdoctoral fellows to come to work for one rather than for one's competitor in a neighboring department. They may also include well-equipped offices and administrative spaces, as well as the provision of courses for students to capture the best candidates for later collaborations. Here, however, I focus on the stakes that matter most in redefining how scholarly communication is undertaken by researchers.

Thus the desire to receive recognition from other researchers may push some to commit fraud, the desire to obtain financial rewards may motivate gaming, and the desire to keep one's job may drive one to play the evaluation game (Crocker & Cooper, 2011; Liu & Chen, 2018). Of course, it is not the case that these categories are always distinct. However, the desire to receive recognition from the top scholars in a discipline will not motivate a researcher to publish an article in a local journal. Indeed, such recognition is practically possible only if research results are published in the most recognized journals in the discipline. On the other hand, a publication in a local journal may be incentivized by a reward system that counts such publications, or by the evaluation system used at a given scientific institution, when such a publication in a local journal allows researchers to meet the evaluation requirements set for them.

The geopolitical context in which researchers work produces variations in the types of stakes in the game, but also, more importantly, in the ways in which

these resources are won. In other words, the historical context in which researchers operate in a particular country causes them to perceive the value of a given
stake in specific ways, as well as to evaluate the forms of play for given stakes
in different ways. Thus the stakes of the game vary depending on numerous
conditions, with players having different starting points, for instance, as a function of the funding their institutions provide for research or in the terms of their
employment. In a country where the majority of papers are published in the
local language and where monographs are considered to be the most important
publication channel (e.g., in the humanities), achieving international recognition is neither easy if one does not publish in English nor all that desirable for a
researcher whose primary motivation is not scientific discovery but retaining his
or her job. This is due to the fact that publishing in local publication channels
allows researchers to get recognized by others in their country and thus to meet
the evaluation system's guidelines. Because such researchers are doing exactly
the same thing as many established scholars in their country, recognizing local
publications as of great scientific value is in their interest and reproduces the
conditions under which they maintain their power. This is the situation in Central
and Eastern Europe within the humanities and social sciences and beyond.
Geopolitical context therefore makes even the stakes appear similar (keeping
one's job, intrinsic satisfaction, recognition); yet the ways in which they are
played for through scholarly communication vary across countries. And this variation is conditioned mainly by the degree of metricization and economization of
science and higher education.

Institutions and Their Managers

Academia is founded on institutions and the researchers, managers, and administrative and technical staff working in them. When I refer to institutions as key
players in the evaluation game, I am speaking primarily of senior officials, that is
rectors, chancellors, and deans, who sometimes hide their motivations and actions
under declarations such as "it is our university's wish," "our institution is trying
to enter the TOP10 best universities in the world," etc. Within scholarly communication, higher education institutions, basic and applied research institutes and
academies of science play a fundamental role that focuses on creating and maintaining working conditions for researchers by providing them with facilities, infrastructures, and financial and human resources. These institutions can also directly
engage in the circulation of scholarly communication when one of their tasks is
to publish scientific papers, as university presses do. However, due to their role in
the scholarly communication system, I classify all publishing houses, including
university presses, as a separate type of player.

The power relations that shape how institutions work are global as well as national and local (Marginson, 2018). At the global level, there is competition for financial resources (e.g., international grants), human resources (e.g., attracting the best scholars), and for the best foreign students. Because of global university rankings, institutions, regardless of national circumstances, place great emphasis on their international visibility and position in various rankings (Yudkevich et al., 2016). At the national level, the power relations in which institutions are entangled that shape how they play the evaluation game are related chiefly to funding mechanisms. Institutions need funding to achieve their primary goals (teaching, research, and development). Funds can come from tuition paying students, donors, state or federal funding, and from grant-making institutions. It is worth noting that the geopolitical location of institutions and the characteristics of different national science and higher education systems are of great importance here. Countries in which the overwhelming majority of students are educated in public universities and do not pay tuition fees (such as, for instance, Poland, the Czech Republic, or Germany) produce a completely different network of power relations than those in which students pay tuition fees (as in the United States or UK) and in which leading universities are not public, but private institutions. Also crucial is whether the university funding system is based on an algorithm that takes into account the number of students and faculty or whether it is fully or partially dependent on the results of a national research evaluation system.

Managers of higher education and research institutions are motivated to support researchers in their scholarly communication practices, as this helps to achieve institutional goals. The race for ever better positions in the rankings intertwines the processes of the metricization and economization of academia. By changing the structures of universities, modifying the ways in which researchers are hired, and incentivizing publications in top-tier journals, institutions seek to orient themselves according to metrics. Such steps, where they result in high positions in the rankings, have the potential to increase profits and the return on investment, for example, through the employment of top publishing researchers. Metricization and economization are two sides of the same mindset that is based on the drive to accumulate values such as prestige, resources, and capital.

Information from research evaluation systems is limited to managerial discretion and may be utilized either in the justification of reforms or as a basis for various incentive systems (Gläser et al., 2010). In the current globalized and metricized system of science and higher education, managers ensure that their institutions are not only ranked but also that they continually improve their position. As I showed in Chapter 4, university rankings are based largely on the products of scholarly communication. Therefore, it makes institutions highly motivated to have staff who can publish the best articles and monographs, from the perspective of rankings and research evaluation. Moreover, if an institution is in a country where there is a national research evaluation system, then rectors, deans, and directors of institutes can motivate (e.g., through a system of incentives or periodic evaluation of employees) their staff to publish research in such a way as to meet the expectations of the national evaluation system.

Importantly, there may be instances in which the goals and criteria of national systems and global rankings diverge. Just as scholars face dilemmas produced by disciplinary and institutional loyalties, institutional administrators may also confront the following dilemma: Is the outcome of an evaluation in a national system more important than a university's global ranking position? The answer to this question clearly depends on many factors, notably the goals of the institution, the manager's personality, staff, and resources. However, the changing terms and conditions of rankings and national evaluation make it a difficult task to determine how to achieve an institution's goals.

Publishers and Editors

In publication-oriented academia, publishers and editors are not only key players in the evaluation game but also crucial actors and gatekeepers who sustain the academic publishing and scholarly communication system (Zsindely et al., 1982). The role of publishers, both commercial and noncommercial, is to produce and maintain the infrastructure (scholarly journals, book series, and repositories of documents and data) of scholarly communication, to peer review–submitted manuscripts, and, most importantly, to publish papers. Editors create, develop, and maintain the scholarly journals that publish research findings and reproduce scholarly networks. The publication of research not only makes it possible to read the results of often longstanding research but also to use it at a later date as well as to verify the analyses carried out.

Like all corporations, large commercial publishers operate for profit. Thus, all changes in their scholarly communication practices (e.g., the establishment of new journals) are motivated by a desire to increase their profits, which in fact derives from the economization of academia and makes this process of economization stronger. In scholarly communication, however, there are still many noncommercial publishers like university presses and publishing houses that belong to learned societies or scientific associations whose main goal is to communicate the research results of their employees or members. However, regardless of whether a publishing house is motivated primarily by the desire for profit or the communication of research results, the way in which it functions remains strongly linked to other actors in the science and higher education sector who are operating in a metricized context. In particular, publishers and editors have to deal with changes in scientific policies at the global and national levels and with the rules for indexing journals and books in the major citation indexes.

The power relations in which publishers and editors are entangled are not, at first glance, directly related to research evaluation systems. However, it appears that they are highly sensitive to transformations in both national-level science policy and the policies of individual institutions. In academic publishing, large actors, the so-called oligopoly of academic publishers, play an increasingly important role in imposing their rules of the game on the publishing market. However, even such publishers must adapt to global public policies. For instance, in 2018, a group of national research funders from twelve European countries established cOAlition S to provide immediate open access. In addition, they set out to make unrestricted use of open access a requirement for all published research – publications and data – funded by the signatories and any funders from around the globe who wished to join (Korytkowski & Kulczycki, 2021). One outcome of this initiative has been to oblige publishers to transform their businesses in order to meet the requirements of this transnational policy.

The strongest power relationship, which is usually asymmetrical, is that between publishers (especially noncommercials, university publishers, or journals of scientific societies) and international citation indexes and bibliographic databases. For most publishers, and therefore for editors who manage journals or book series, being indexed in WoS or Scopus and having a calculated JIF is the be-all and end-all. A newly created journal can operate on the publishing market for two to four years at most without any calculated bibliometric indices, relying on the enthusiasm of authors and editors. After this period, however, even the most ambitious editor and the most understanding publisher will see an exodus of authors, for whom publishing in a journal that is not promising (and a multi-year journal with no Impact Factor in almost any scientific field will be considered such) is a waste of a publication that could otherwise help them win stakes in the game. Moreover, in non-English–speaking countries, in the humanities, social sciences, and other fields, there is, of course, a publishing market that is only to a very small extent visible in international databases (Bocanegra-Valle, 2019; Salager-Meyer, 2015). However, this does not mean that publishers and editors in those countries are not bound by evaluative power relations based on bibliometric indicators. If a country has a list of journals recognized in research evaluation or in a financial reward system (as is the case in China, Denmark, Finland, Norway, Italy, and Turkey, among others), the journal must be on this list so that local authors do not stop publishing in it.

Publishers and editors have to constantly adapt to the regulations of science policy and national science evaluation systems in order to maintain their position in the publishing market. This applies especially to local journals published by regional universities or learned societies. Science policy and evaluation systems rely heavily on bibliometric indicators based on citations. Thus, it becomes the role of editors to recruit the authors of the best texts that are most likely to be cited. Such authors are attracted by, among other things, other authors who have already published in the journal, the composition of the journal's editorial board (the more key figures in the field, the greater the chance that top scientists will want to publish there), and the fact of having high bibliometric indicators (especially JIF), which allow researchers to maintain their position through various forms of evaluation game.

Policy Makers

Designers of research evaluation regimes and policy makers involved in global, national, or local policy are also sometimes players in the evaluation game. The playing of the game starts when policy makers implement evaluation regimes. However, once started, the game transforms research practices in academia and this requires a response by policy makers. They must adjust the rules to the transformations and update policy and political goals that are linked to, among other things, global university rankings or an increase in the national scientific system's competitiveness. The manner in which evaluation systems are designed depends strongly on both the scientific position of a country and the associated goals to be achieved. Thus English-speaking countries will not aim to increase the proportion of publications in English, as most countries in Central and Eastern Europe or Asia do.

Behind the implementation of any research evaluation system there is a political motive. This may, for instance, be to increase transparency in public spending or to raise the quality of research conducted at universities as expressed in the number of publications in top-tier journals. The creators of these systems, along with the policy makers who maintain them are aware that every such system will be played, and that managers and researchers will try to play the evaluation game created by them. Therefore, these systems are constantly being improved: The rules for counting publications are changed and new databases and bibliometric indicators used. Each subsequent change makes it necessary to recommunicate the evaluation rules and to respond to emerging problems. For example, researchers may start publishing in low-quality journals because the evaluation system recognizes them and, moreover, treats them in the same way as publications in very good publication channels. When policy makers take note of the negative consequences of their actions, they try to maintain their position by imposing new rules. This is often justified by a new context and the achievement of previously stated goals. As I experienced during the course of several years in which I served as a policy advisor, in such moments, policy makers will start by emphasizing that the previous provisions had some imperfections. Nonetheless, they will go on to say, it was a step in the desired direction in that it promoted the kind of scholarly communication which is desirable from the perspective of science policy within the country or institution. In such cases, evaluation systems can be supported, for example, by excellence initiatives (such as the Russian and Chinese projects described in Chapter 4), which are also aimed at achieving science policy goals as expressed through positions in various global rankings.

Although policy makers, in collaboration with researchers, are largely responsible for creating evaluation systems, they also play the evaluation game. This is because evaluation systems have set goals (e.g., increasing the internationalization of scientific publications) and, at the same time, are constantly being criticized by the scientific community that is supposed to meet these goals. Therefore, policy makers need to keep evaluation systems in a process of continual change in order to maintain their power position.

* * *

What types of evaluation game can players game? All players, whether researchers or institutions, can play the same types of game. Moreover, they may frequently be playing on the same team. For example, this occurs when institutions use the principles of the national evaluation of institutions to evaluate their staff, and researchers, in fulfilling institutional requirements, publish their papers in journals that count for the evaluation, but are not necessarily key journals from the perspective of a particular scientific discipline. Players can play to increase the number of publications that will be counted in the evaluation or global university rankings. Alternatively, where we are referring to play by publishers or editors, they may try to increase the number of citations of papers published by researchers from their institution or journal. There is one further critical question that must be answered with respect to players of the evaluation game. It stems from the argument developed in this book so far. That is, do players who work in countries whose current evaluation regimes have grown from socialist models (described in Chapter 3) play the evaluation game differently than those whose evaluation regimes were based on the neoliberal model from the outset? Australia and Poland have somewhat similar national systems of research evaluation. To put it in a nutshell, Australia has a model based on neoliberal British approaches. Poland has a similar model that was, however, first implemented during the transition away from a centrally planned system.

Thirty years have passed since that time. Do reactions to research evaluation regimes in Australia and Poland look similar today? In other words, do Australian researchers play the evaluation game similar to their Polish counterparts or not? When one looks at studies on the effects of research evaluation systems in these two countries, one finds that researchers play the evaluation game in a very similar fashion (Butler, 2003a; Hammarfelt & Haddow, 2018; Kulczycki, Rozkosz, & Drabek,

2019; Kulczycki et al., 2020; Woelert & McKenzie, 2018). What differs significantly across these science systems is the level of "trust in experts" (high in Australia and the UK and low in Poland or Russia) and "trust in metrics" as a means of salvation from experts (high in Poland and Russia and low in Australia and the UK). This shows that it is not so much a question of how players play the game differently, but rather, of how historical context has shaped the evaluation game, and how, as a result, players perceive evaluation systems. A Polish dean would implement the national research evaluation system's rules but would at the same time admit that this was a desirable state of affairs. At long last, everything is clear because it is expressed in metrics and algorithms, deans know what to do, and no one can claim that evaluation results are the dean's fault, because it is simply a question of how the numbers add up. In addition to trust in metrics, which in my opinion in Poland and Russia has Soviet roots that grow out of Russia's imperial heritage, one can also observe that in countries like Poland, loyalty to the institution plays a much more important role than loyalty to the discipline. Researchers in Poland are far more likely to follow a path that is consistent with the demands of their institution, or their perceptions of those demands, even when this is not consistent with disciplinary demands. They will opt for this course rather than following a path that is consistent with disciplinary loyalty but is somehow either inconsistent with institutional loyalty or simply requires more effort.

The three cases that I describe in Chapter 6 illustrate how trust in metrics is greater in countries whose research evaluation systems have a longer history than the British or Australian regimes. In addition, the case studies also show how strong loyalty to institutions is a product of socialist heritage. Further, as I argue, deploying the concept of playing the evaluation game allows us to illuminate different aspects of the game than does the concept of gaming. This is because people game in the same way everywhere, but their reactions to evaluation regimes differ when evaluation systems have socialist as opposed to capitalist roots.

6

Playing the Evaluation Game

Occasionally, one becomes a player of the evaluation game without having been invited to join it by anyone. In the story I unfurl in this chapter, all types of players appear: a publisher, researchers, university managers, and policy makers. All of these figures were drawn in to play a joint evaluation game, centered around a scholarly journal. This was an open-access journal published by a Greek association which, without asking for or seeking to, became an unwitting beneficiary of the evaluation game. At the same time, it became one of the most talked-about examples of the unpredictable effects of the research evaluation system in Poland.

In 2018, the *European Research Studies Journal* ceased to be indexed by the Scopus database. During the nine previous years, the editor-in-chief had managed to build a community of scholars around the journal who published in it and increased its bibliometric indicator values. But in 2018, due to "publication concerns," Elsevier (owner of Scopus) decided to stop indexing the journal in their database. As it was also not indexed in the Journal Citation Reports, it also had no Impact Factor. This means that for many researchers, after it was dropped from Scopus, the journal lost any bibliometric legitimacy. For most journals, such a loss marks the beginning of the end, which fact gets signaled by a dramatic exodus of authors to other journals. However, bibliometric indicators for the *European Research Studies Journal,* calculated on the basis of Scopus data from 2017 (the last full year in which the journal was indexed) were used (alongside bibliometric data for another 30,000 journals) by Polish policy makers to create the Polish Journal Ranking in 2019, a key technology of the Polish evaluation regime. Through an informed-expert assessment, the journal was assigned 100 out of 200 possible points. In reaching this point tier, the journal became a very attractive publication venue for Polish researchers, especially from the social sciences. Polish evaluation rules state that the number of points assigned does not get divided among co-authors from different institutions when a journal is credited with at least

100 points. Thus, if three researchers from three different Polish universities publish an article together, then each of them will generate 100 points for their university.

A new Polish Journal Ranking that included the *European Research Studies Journal* was published on July 31, 2019. Before that date, few Polish authors had published in the journal. For example, in Issue 1 in 2019, out of seventeen articles, only one was by a Polish author; similarly in the next issue, one out of twenty-three articles was by a Polish researcher. But from the publication of the ranking onward, the percentage of Polish authors and the total number of publications by the journal began to increase dramatically (Wroński, 2021). In the last issue of 2019, out of thirty-seven articles, as many as 73% had Polish authors. In 2020, in addition to its four regular issues, three special issues were published, which were basically entirely "taken over" by Polish authors: In the first issue, out of seventy-six articles, seventy-four (97%) had at least one Polish author. In the second and third issues, this share was 98% percent. In the last special issue in 2020, two-thirds of the articles were authored by researchers from the Military University of Technology in Poland, mostly from one department. One researcher from this university co-authored as many as six articles in this special issue.

Although the journal lost its bibliometric legitimacy, it received not only the Polish evaluation regime's legitimation but was even distinguished by being assigned 100 points as an outlet worth publishing in. Given this situation, the journal opened fully to Polish authors and allowed itself to be taken over. After all, each publication is followed by a respectable sum of money, as the journal operates on the Gold Open Access model: It is the researchers (or, more precisely, their institutions) who pay for the publication of articles that it accepts. The cost of a single publication, taking into account the financing of Polish science, is not insignificant and ranges from 600 EUR to 1,000 EUR, depending on the discounts granted. Discounts are varied and depend on participation in a conference in which the journal is a partner. If one participated in the conference, one could publish one's conference paper in the journal, after a positive review and payment of the fee.

It should be noted that this increase in the number of Polish authors in the journal is not solely the result of the desire of Polish researchers to obtain points for the evaluation of Polish universities. This is because most researchers are probably aware of the fact that such publications are not in accordance with international disciplinary loyalty. Such a situation would not have occurred if it had not been fully supported by the institutions employing these researchers and the journal which flexibly adapted to the changing situation. The journal's editorial board includes several Polish researchers, including university managers, notably the rector of a Polish university who headed the organizing committee of the 2020 and 2021 conferences co-organized by the journal's publisher. Polish universities have started organizing conferences in partnership with the journal's publisher, and one of the Call for Papers includes information (in the headline itself) about the "possibility of publishing for 100 points." Such events are organized with the support of universities and deans. Presentations from these conferences are subsequently published in the *European Research Studies Journal*; with the costs of open access covered by universities. In this way, deans can improve their score in the upcoming national evaluation exercise. The cost of a few thousand euros does not seem that significant when compared to the consequences that may befall universities that score low in the evaluation.

Universities promote these publications on their websites and in social media, as for many departments, publication for 100 points is a reason for "evaluation celebration." Moreover, universities often provide, in addition to information about the authors, the number of points they "brought to the university." They also advertise the scores obtained in bogus bibliometric indicators which the journal glorifies. For example, the Three-Year Impact Factor by Thomson Reuters, while the Impact Factor is either Two Years or Five Years and, since 2016, Thomson Reuters no longer owns the database used to calculate this indicator.

As noted, the publisher has adapted to the changed situation and allowed Polish authors to take over the journal. Here one can observe a "mark" that signals the impact of the Polish research evaluation regime. Thus participants of a conference organized in Poland can publish their articles in this Greek journal for 720 EUR, but only if the article has one or two authors. For each additional author, they need to pay an extra 100 EUR. In the Gold Open Access business model, researchers pay per article or per number of pages, but not per number of authors. In this case, however, the publisher has adapted its business model to Polish evaluation rules. For instance, six scientists from six Polish institutions could get 100 points each for their institutions for "only 720 EUR." But for such an article, one has to pay an extra 400 EUR for the "excess" authors.

The case of this journal has become famous in Poland: It was discussed not only by researchers in social media and blogs but was also described in detail in the most important academic magazine, *Forum Akademickie*. However, the Ministry of Education and Science has not decided to remove the journal from the ranking. What arguments were used to justify this non-action? First of all, there are more than 30,000 journals in this ranking, and it can happen that the quality of a journal slides from normal to questionable. Second, all the publications in this journal by Polish authors are in English. In a strange way, this fact fulfills policy makers' goal of internationalizing Polish science and providing broad access to Polish scientific results. Officially, policy makers explain their decision not to remove the journal by a desire to avoiding changing the rules of the game during the game (the term "game" is used explicitly by them). This is despite the fact that the rules are constantly being changed anyway.

What then is the real issue? Could it be a fear of lawsuits (research evaluation in Poland is an administrative procedure), or perhaps the high-level power relations at work in the ministry, since large Polish universities are involved in this evaluation game? Or perhaps the Ministry sees something in the journal that is overlooked by those researchers who consider publishing there a waste of taxpayers' money?

* * *

In this chapter, I utilize the theoretical framework developed in the book to describe three representative fields in which the playing of the evaluation game takes place. These are "questionable academia," "following the metrics," and "adjusting publication patterns." I have chosen to focus on these areas for two reasons. First, they represent some of the most pressing problems in contemporary scholarly communication, and yet at the same time, are under-theorized in research evaluation and higher education studies. Second, I have spent the last decade investigating the transformations of scholarly communication in these fields that has resulted from the metricization and economization of science systems. Therefore, on the basis of both my own research and that of other scholars, I have a story to tell about the impact of publication metrics on scholarly communication.

The case of the *European Research Studies Journal* focuses all three areas within a single lens: one is dealing here with questionable journals, university managers "following of the metrics, and a change in the publishing practices of Polish researchers. However, each of these issues deserves its own story to aid understanding of the transformations in research communication caused by the metricization and economization of science. The cases I discuss in relation to the first field allow me to explain why publishing in predatory journals (or publishing in low-quality journals) can be considered not only as cheating or gaming but also as a form of playing the evaluation game that results from strong "institutional loyalty." Through the second field, I highlight the fact that researchers in different countries game the system in the same way because their goal is usually to maximize profits, but when it comes to playing the evaluation game, they play differently depending on the geopolitical and evaluation contexts. With the cases included in the last field, I draw attention to the transformation of scholarly communication that is visible at the level of publication patterns, that is, transformations in where scholars publish and what channels, languages, topics, and publication types they choose. The cases presented in each field are structured in a similar way. First, I describe the research practice in question (e.g., publishing in predatory journals). I do so without using the term "evaluation game," but relying rather on such terms as are current in the literature. On this basis, I present existing explanations of the given practice and its transformation. I then move on to unpack the practice using a framework based on the evaluation game. Thus I characterize players across three intersecting planes (global, national, and local) and then clarify what the playing of the evaluation game consists in, before finally summarizing play in this field in a changing geopolitical context.

6.1 Questionable Academic Practices

One of the most discussed topics in studies on scholarly communication across all fields of science is predatory publishing. In addition, ever greater concern is being directed toward predatory conferences and the bogus metrics that are used by low-quality journals to pose as prestigious outlets with reputable bibliometric indicators. For many researchers, the term "predatory" is inaccurate and misleading, because it is not always the case that such journals or conferences are oriented toward financial profit (this is a key feature of being a predatory publication channel). Nor is it always true that articles or conference materials published through such channels are of low quality. Thus it may happen that a researcher who is unaware of a journal's rank sends it an article whose research findings are adequate. Moreover, as a recent study has shown (Kulczycki et al., 2021a), articles published in predatory journals are cited by reputable journals (including those with a Journal Impact Factor) and those citations are in their overwhelming majority neutral. One might claim that some of the worthless articles published in predatory journals may still leak into mainstream literature as legitimized by WoS, or that some papers, especially from the Global South and Global East published in predatory journals are somehow important for developing legitimate scholarly discussion. For these reasons, I prefer to use the term "questionable" over "predatory" to name scholarly communication practices that are concentrated in three main areas in which the evaluation game is played. These are: editing and publishing in predatory journals (and book series), organizing and presenting at predatory conferences, and using bogus metrics.

Significant interest in questionable academia began in 2012 when Jeffrey Beall (2012), an American librarian, published a *Nature* paper in which he warned against open-access publishers who are ready to publish any piece as long as they receive an author's fee. A year later, he added: "The predatory publishers are a false front; they intend to deceive, and sometimes honest scholars and honest institutions fall into their traps" (Beall, 2013, p. 14). Beall also ran the blog "Scholarly Open Access," where he published what for many scholars were controversial lists he had drawn up of predatory journals and publishing houses. The lists quickly provided the basis for numerous studies demonstrating, sometimes in a spectacular way, the inability (or unwillingness) of predatory journals to verify the content they publish (Bohannon, 2013; Sorokowski et al., 2017). Research published on the topic has covered the increase in the number of articles published in predatory journals over time (Perlin et al., 2018) as well as the countries of origin of the authors who publish in such outlets (Xia et al., 2015) and the share of predatory journals within a given discipline (Noga-Styron et al., 2017; Yan et al., 2018). Despite Beall's blog closing down and his two lists being discontinued in 2017, his impact on the debate concerning predatory journals and his linking of predatory publishing with open-access models is still tangible today (Krawczyk & Kulczycki, 2021a). However, even before the discussion on predatory publishing, numerous papers had aimed to show the weakness of the peer review process at certain journals or conferences. One of the most famous of these was Alan Sokal's (1996) hoax paper aimed at parodying postmodern jargon published in prestigious journals in the humanities.

In 2017, together with several colleagues, I published the results of a sting operation commonly known as "Dr. Fraud" (Sorokowski et al., 2017). In 2015, we had created a fictional researcher, Dr. Anna O. Szust (in Polish "Oszust" means a "fraud") and on her behalf, had applied to the editorial boards of 360 so-called predatory journals, of which 48 accepted her and 4 made her editor in chief. Despite this disclosure, Dr. Szust continues to "be used" not only as a journal editor but also as a member of the international advisory boards of many conferences – even several years after the publication of the sting operation results. For many years now, multiple texts have circulated warning against predatory or questionable conferences which are co-organized by researchers such as Dr. Fraud (Bowman, 2014; Koçak, 2020; Lang et al., 2018; McCrostie, 2018; Pecorari, 2021; Teixeira da Silva et al., 2017). Such conferences are characterized by their use of hijacked photos and biographies of researchers from prestigious institutions, the use of names that resemble those of long-standing, reputable conferences, and the hosting of low-quality academic meetings that gather researchers from practically all scientific fields in a single event. Many researchers – both early career and senior – might therefore be tricked into attending such a conference.

The harm produced by predatory journals is caused by the lack of scientific control over the content they publish. Editors of such journals might be nonexistent researchers (Sorokowski et al., 2017); an article on bats written by a seven-year-old may get accepted (Martin & Martin, 2016) as might a poorly fabricated study on the development of cancer (Bohannon, 2013). For this reason, predatory journals have begun to be seen as a serious threat to academic discourse as a whole (Beall, 2018; Moher et al., 2017). Furthermore, predatory journals often use bogus metrics such as the numerous "variations" of the impact factor (Dadkhah et al., 2017) which are calculated by incompetent companies or, in actual fact, not calculated at all. Editors may simply indicate on journals' webpages that journals have some sort of important impact factor equal to 3.421.

In terms of the factors producing questionable academia, these are often identified as the publish or perish research culture, the proliferation of open-access
models in academic publishing, and various national evaluation frameworks that
generate pressure to publish ever more (Lakhotia, 2017; Omobowale et al., 2014).
In publication-oriented academia, the pressure to publish, as I have shown, is an
inevitable effect of metricization and economization. This pressure was recognized
by various commercial publishers and entrepreneurs that had previously been unin-
terested in academic publishing. Taking advantage of the opportunities offered by
online publishing, many new journals began to be established in which authors
paid to publish. Increasingly, these journals were founded not by learned societies
or universities, as had previously been the case, but by profit-oriented publishing
companies. This kind of set up is not always a bad thing, as it relieves researchers
of difficult and tedious technical work. At the same time, however, this aspect of
the academic publishing market has become fertile ground for the development of
predatory journals.

The pressure to publish has a specific and material geopolitical context.
Peripheral researchers are hindered in a number of ways: shortcomings in the available research resources, insufficient English language skills, and a lack of adequate
socialization enabling them to navigate the global publishing market effectively
(Kurt, 2018; Önder & Erdil, 2017). Frequently, those institutions subject to these
pressures do not themselves have the right tools or competent experts for assessing the quality of scientific publications which are required by pressure-bearing
institutions. Thus, they focus on assessing the number of articles rather than their
quality. As Omobowale et al. (2014) have shown in the case of Nigerian promotion
committees, peripheral institutions are often unable to distinguish a recognized
journal from a predatory one, and, even when they do, they are often guided by
the committee members' affection or dislike of the academic being assessed. The
lack of suitable tools for proper assessment of academics at the peripheries also
has a structural basis. As noted by Ezinwa Nwagwu and Ojemeni (2015), the largest databases indexing scholarly publications omit many journals from peripheral
countries. Further, the cost of accessing the content of the lists that verify the
quality of journals, such as the Journal Citation Reports, may be too expensive for
peripheral institutions (Gutierrez et al., 2015). However, the main problem with
explaining the phenomenon of predatory journals in terms of peripheral "gaming
the metrics" is that it overestimates the role of only one of the factors shaping
the modern academia. For instance, the core problem with political science in
Argentina, as described by Rodriguez Medina (2014), is not the overly strict publication requirements imposed by institutions, but rather the overburdening of
academics with heavy teaching loads. Similarly, for Belarusian scientists, publications in international journals do not even count within institutional assessment

(Warczok & Zarycki, 2016), thus they lack institutional incentives that could encourage them to publish in predatory journals.

Numerous studies have shown that peripheries (sometimes labelled as "developing countries") are more profoundly affected by predatory publishing than the Global North. Xia et al. (2015) found that the three most frequent country affiliations of authors publishing in predatory journals were India, Nigeria, and Pakistan. Demir (2018) identified India, Nigeria, and Turkey. Although the countries identified may differ depending on each study's methodology, almost all studies seem to agree that they are overwhelmingly peripheral countries (Erfanmanesh & Pourhossein, 2017; Macháček & Srholec, 2017; Shen & Björk, 2015). Moreover, most predatory journals operate in peripheral countries (Shen & Björk, 2015). Perlin et al. (2018) found that in Argentina, scholars who obtained their Ph.Ds locally were more likely to publish in predatory journals than those who obtained their PhDs abroad. Nevertheless, predatory publishing is present not only in the peripheries but also in the center. For example, in a study on predatory journals focusing on tourism and hospitality, Alrawadieh (2018) found that most authors had affiliations in the United States, followed by Nigeria, Taiwan, Malaysia, and Turkey. Moreover, researchers from the United States constitute the highest share of participants in questionable conferences organized by the OMICS International group (Kulczycki et al., 2022). However, it is important to see such numbers in the context of total publication output and the number of researchers in a country. According to the UNESCO Science Report (Schlegel, 2015), in 2009, there were twenty times more researchers working in the United States than in Turkey and forty times more than in Malaysia.

Publishing in predatory journals or attendance of predatory conferences attracts public attention when researchers win awards for publication in low-quality journals or when a sting operation reveal that it possible to publish literally anything (even nonsense text) in a journal. The Indonesian system for measuring performance in science was intended as a means of giving recognition to Indonesian researchers, but with time, it allowed numerous researchers to become top scorers in the national ranking thanks to the fact of publishing in low-quality journals (Rochmyaningsih, 2019). Similarly, there is concern within the academic community in South Africa about the way that payouts push professors toward predatory journals (Hedding, 2019). Public interest in researchers' attendance of questionable conferences is also linked with the fact that this is a way of accumulating undeserved profits (McCrostie, 2018). In 2019, South Korea announced a new policy, designed to reduce researcher participation in "weak" conferences (Zastrow, 2019), a phenomenon that had previously been confirmed by analysis and a sting operation (Chung, 2018).

Questionable academia, however, is not only the field in which gaming takes place. It is also a product of various global, national, and local types of evaluation game. Companies and editors do not always establish new journals in order to make money on article process charges, and researchers do not always publish in predatory journals in order to get monetary rewards. Furthermore, it is not always the case that policy makers design evaluation regimes that count publications in predatory journals because of ignorance of the rules and of scholarly communication's ethos. As the case presented at the beginning of this chapter showed, sometimes questionable academia is a systemic effect of research evaluation regimes. Thus low-quality journals are founded not simply to earn money but also in order to publish papers that count in research evaluation regimes. Various Polish journals, especially in the humanities and social sciences, were founded because it was easier – in terms of the requirements of the national research evaluation regime – for researchers to establish a new journal and publish their work in it than to try and publish an article in a reputable journal. Such Polish journals and their editors have strictly followed the Polish Journal Ranking criteria rather than the discipline-specific practices of scholarly communication. Although these journals do not take article processing charges (which is often a key, but on its own not sufficient rationale for determining whether a journal is predatory) they can be called questionable because their publication goals and methods of soliciting and reviewing articles are of low quality. Therefore, when the Polish Journal Rankings guidelines gave more points to journals that had a high percentage of foreign editorial board members, editors – seeing this as the simplest solution for following the rules – asked Polish members to leave the editorial board, rather than inviting foreign scientists. This ostensible internationalization was, as confirmed by interviews with editors (Kulczycki, Rozkosz, & Drabek, 2019), caused by observance of the research evaluation system's rules. The same is true for researchers who publish in questionable journals. They might opt to do so not for monetary gain, but because science policy in their country is focused on international publications. Thus an article in any journal that is published outside their country is considered more desirable than one published in a local journal and in the local language.

The debate on predatory publishing focuses almost entirely on English-language journals published in non-English–speaking countries. Solutions such as Beall's List judge journals on their editorial practices and divide the world into *goodies*, published mostly in central countries in English, and *baddies*, published mostly in English in the Global South or Global East. This is a false dichotomy. There are many bad journals with aggressive business models in the center and many good journals published in English and – primarily – local languages at the edges. And many editorially reputable journals from large commercial publishers have predatory business models.

To get away from an oversimplified view of scholarly publishing, Franciszek Krawczyk and I have coined the term "mislocated centre of scholarly communication" (Krawczyk & Kulczycki, 2021b) to describe and criticize the role some publication channels play in the (semi)peripheries without condemning scholars who publish in them or accusing publishers of bad intentions. Mislocated centers emerge at (semi)peripheries because the center is an essential source of legitimization in the peripheries, and because there is considerable uncertainty in the peripheries over what is central and what is not. The key, however, is that researchers publish in such channels because they are counted in research evaluation regimes at (semi)peripheries. They are counted because they are perceived as connected with the central countries and institutions.

In Poland, there is a company that draws up its own "journal ranking" and its own bogus bibliometric indicator which is used by various journals to demonstrate their relevance. This company was involved in the governmental project to build a "Polish Impact Factor" which was to show the citation impact of Polish journals on Polish journals. This project was misguided, and the government eventually withdrew from it. It was problematic both from a bibliometric point of view (in that it took into account only Polish journals and not book publications) and in terms of its promotion of the internationalization of Polish science. However, the company gained some legitimacy in the Polish scientific community through its cooperation with the government, although outside Poland, it is identified as a shady company that creates bogus journal rankings and metrics.

Sometimes researchers publish in questionable journals because the credibility of those journals is validated by the various lists of journals used in national research evaluation systems (Pölönen et al., 2020), which include journals identified as predatory. Moreover, numerous researchers attend questionable conferences organized abroad because they can achieve better evaluation results for presentations at international conferences. When institutions do not provide enough resources for researchers to do thorough work on generating scientific results, and when evaluation criteria are frequently redefined, scientists may respond in two ways. They may come to value institutional loyalty (fulfilling their institution's requirements) more than disciplinary loyalty (global science publication venues to which one should submit manuscripts). This is because it is by following institutional requirements that they are able keep their jobs. Thus, if national evaluation regimes legitimize publishing in questionable journals and if it is easier to publish in them than in more reputable journals, then researchers will publish in them. This is how they can meet evaluation guidelines at the lowest cost.

One might ask why academics working at such institutions are not driven by disciplinary loyalty. In Poland, there is today a deep crisis of socialization within disciplines, especially in the humanities and social sciences. This is a product of the massive expansion of higher education in the first two decades after 1989, during the transition from the socialist system (Antonowicz et al., 2020; Shaw, 2019). Researchers were given the choice of either dramatically increasing their teaching loads and thus achieving higher salaries, or, if they chose not to do so, of receiving salaries that did not allow them to support their families. The fact that many chose the first option had a drastic impact on Polish universities' research mission and on the international visibility of research. It also meant that institutional loyalty, which allowed for higher salaries when additional courses were assigned to a researcher, played a greater role than disciplinary loyalty.

It is commonly asserted that questionable academia is mainly the domain of the Global South and of countries such as India, Turkey, or Pakistan. However, in these countries, as well as in Poland and Italy, and indeed most countries in which researchers publish in questionable journals (Kulczycki et al., 2021a), this kind of publishing or participation in questionable conferences is primarily a way of maintaining one's status by producing an adequate number of foreign publications and presentations. Moreover, policy makers are usually aware that national-level evaluation requirements (both for universities and individual researchers) must be achievable for the majority of institutions and researchers. This is due to the fact that the quality of science in their countries, as measured by publications in international journals, is not high, and individual institutions differ little from each other from the perspective of international rankings. At the same time, these policy makers want to follow the logic of internationalization and increase the number of publications in journals included in global university rankings. Thus in the end, only "foreign" papers actually count. The question here, in relation to Poland, is why do Polish policy makers not remove such journals from the Polish Journal Ranking? After years on the "dark side" of evaluation power, serving as a policy advisor, my understanding is that the actual reason for this is that they are afraid that universities would take the ministry to court for radically changing the rules of the "game" during the game. Numerous institutions have already invested funds (i.e., paid APCs) to get better evaluation results (as in the case of deans supporting publishing in the *European Research Studies Journal* described earlier). It is therefore in the interests of such institutions to keep questionable journals in the rankings. By contrast, policy makers' interest – their personal interest, not the interest of the values they represent – is peace of mind and an absence of resentment by stakeholders.

The case of the *European Research Studies Journal* is not unique, even in Poland. During the 2013 research evaluation exercise, a Polish journal started to publish hundreds of papers every two months, adding them on CD-ROMs to the paper version of journal in which only a few articles appeared. Several institutions built their research portfolio for evaluation purposes primarily on the basis of this journal. However, this journal differed in a significant way from the Greek one: It did not charge for publication, so the fact of publishing so many articles in one issue as a CD supplement only served researchers and institutions in the context of the upcoming evaluation.

When researchers follow evaluation rules that count publications in questionable journals, they are not gaming the system, they are playing the evaluation game by minimizing effort in order to meet evaluation expectations. They publish in predatory journals to give the institution what it expects from them and get peace of mind. This is also a good scenario from a university's perspective: They receive the relevant (i.e., eligible) publications, which are then converted into evaluation results. Finally, from a policy maker's perspective, the situation is also desirable: They create the rules that can be followed by all institutions what (i.e., feasibility of requirements for all institutions) is one of the hidden, unexamined assumptions of the research evaluation system.

6.2 Following the Metrics

With my discussion above, I foregrounded the loyalty to institutions that is manifested by individual researchers. The area I turn to now helps clarify how managers in academia play or support the playing of the evaluation game as a result of the increasing pressure for outcomes and accountability created by the quantification of social practices. In Chapter 1, I referred to Campbell's observation that whenever a metric is used for social decision-making, it will eventually distort and corrupt the social process it is intended to monitor. In light of this, it is worth looking at the practices of policy makers who introduce metrics to monitor and control, sometimes to incentivize desirable (from a policy perspective) research and publication practices. These metrics are followed both by managers of academic institutions and researchers themselves. In line with my overall argument, the practice of following the metrics should be viewed first as playing in step with an evaluation regimes' rules as they are expressed through metrics, and second, as prioritizing the meeting of evaluation expectations over disciplinary loyalty. I focus on two examples in what follows, which center on the establishment of local journals and the reproduction of national regulations at the institutional level.

"Following the metrics" is often presented as a way of gaming the evaluation and scholarly communication systems. Publishers, editors, and researchers can game to increase a journal's impact factor by means of various practices. These start from coercive citations (editors ask authors to add citations from the editor's journal to a submitted article), to the publication of more editorials which are only partially included in the calculation of impact factor, thereby increasing its score, to finally the establishment of so-called citation cartels of researchers who mutually quote articles from selected journals (Chapman et al., 2019; Davis, 2016; Gingras, 2014). For instance, a Turkish journal called *Energy Education Science and Technology* had an Impact Factor of 31.677 and was the seventeenth highest ranked journal in the 2011 edition of the Journal Citation Reports. Even *Science* had a lower impact factor than this journal. Then, it was revealed that the editor of the journal played a game with the impact factor by using self-citations (Umut & Soydal, 2012). Although the journal was withdrawn from the WoS, most of the Turkish scholars publishing in it got tenure and monetary rewards by using articles and citations in it. As a consequence, a fix is currently being sought out that can address the impact factor's weaknesses. The solution to the exploitation of the impact factor as a single metric was to be "altmetrics" (Lin, 2020), a system based on diverse social media, the latter of which can also, however, be partially exploited by bots.

Global actors in science policy such as the European Commission use international citation databases (e.g., Scopus) to tell a story about the current landscape of science and higher education and make political decisions on the basis of it. This prompts national policy makers and institutions to pay even more attention to publications in Scopus and to measures built on its basis that seek to follow global trends. Companies that devise global university rankings, global providers of bibliometric data, and global publishers have been reshaping scholarly communication in a way that makes them the key source of the prestige and visibility sought by universities and other research institutions. This turns the focus of scholarly communication to journals that have "proper metrics." In the case of Elsevier, which is both a global publisher and bibliometric data and metrics provider, one can observe a closed circle. Value (i.e., the high rank of a journal or university) is built on metrics that are calculated by this company on the basis of publications that are frequently its own, given that Elsevier is a key oligarch in the academic publishing market (Larivière et al., 2015b). This is how a single company can create the conditions under which a university is considered valuable. One has to publish in that publisher's journals because they are fully indexed in a database that is used to create the metrics used in university rankings. Moreover, the same metrics are also often used in national science evaluation systems (Sīle et al., 2018). From the perspective of a dean or rector, above and beyond their potential criticisms of such a closed circle, it is rational and reasonable that they implement this pattern of actions to follow the metrics. The same is true with respect to national plans, through which the conditions are created for national policy makers to play the evaluation game. National players and institutions in the science and higher education sector establish their own university rankings and various types of "research excellence initiative" (like the Chinese and Russian ones described in Chapter 4) to make universities perform better in global rankings.

These have such a strong impact on research communication that the metrics used in these initiatives directly copy the criteria from international citation database rankings. At the end of the day, institutional and local players reproduce national solutions that were designed not only for researchers but also for institutions to meet evaluation criteria.

Explanations of why academia is so quick to respond to metrics focus largely on the publish or perish research culture, the need to get grants (as in the Stefan Grimm case discussed in Chapter 2 on economization), and the role of global university rankings in attracting students. These explanations show that the main motivating factor for using and following metrics is to maintain and improve one's position in academia. This pressure created by publication-oriented academia, together with the need to get grants each have their varied geopolitical shades. In an article entitled "If the Indicator Game Is the Answer, Then What Is the Question?" published in a special issue of *Engaging Science, Technology, and Society*, Irwin (2017) calls for the following: "Even if both Britain and Denmark have national research quality assessments, these differ widely in their specifics and in their implications – and variations are evident within any one country. (…) My point is not to downplay international trends and especially career pressures in a globalized academic market. (…) Indicators are now everywhere, but one cannot just take it for granted that their significance is always the same" (Irwin, 2017, pp. 65–66). In other words, the practice of following the metrics is a global phenomenon, just as playing the evaluation game is, but the context in which a particular research evaluation system is implemented distorts academia's response and makes it country-specific.

No matter how well policy makers argue for the use of a given metric, once a metric is established in regulation as a tool for measuring the productivity or impact of an institution or researcher, that metric gets removed from the context in which it was designed. It becomes a "pure metric" devoid of the modalities that gave it meaning, and which supported the arguments of policy makers and research evaluation system developers. For example, if the creators of a research evaluation system place more weight on evaluating universities through the number of articles in journals rather than scholarly books, and if this emphasis is based on the belief that in the current system of scholarly communication, articles are a more important and effective publication channel, then the predicted (by policy makers) result of such regulation will be an increase in the number of articles (a decrease in book publications may not occur at all). Therefore, top-tier journals, which are most often defined by their impact factor, have the highest value for the final outcome of the evaluation in a system. However, the creators of these systems, which are supposed to cover all institutions in a given country, do realize that the counting of articles cannot be limited only to journals with impact factor. This is because they know that in many disciplines, only a small percentage of researchers have such publications, and changing this state of affairs requires a longer period of time. Therefore, articles published in journals other than those with an impact factor can be counted for an institution or researcher. And since it is acceptable to meet evaluation expectations at a lower cost, this opens the field for a new type of evaluation game: the game of establishing local journals.

The game of establishing local journals has a Central and Eastern European flavor. As mentioned previously, it was as early as 1830 that university professors in Russia were obliged to publish a paper every year. According to Galiullina and Ilina (2012), scholarly journals established in the middle of the nineteenth century in Russia were intended by the authorities to have a controlling and stimulating function as a kind of report of each academic community on its research work and on its contribution to the development of state science. Sokolov (2016) argues that this practice was sustained throughout most of late imperial and Soviet history, and that publication requirements survived after the 1990s. This pressure on publications was addressed by universities in Russia in a specific way. Instead of questioning the rules imposed by the authorities, universities started to establish series of journals called "Proceedings of University X." These were funded by university budgets and, of significance here, published papers exclusively by the university's own faculty – with rare exceptions. Sokolov underlines the fact that such journals never reached any of the distribution networks. Their main purpose was not to communicate research but also for universities to follow the metric: the number of articles a professor had to deliver each year. This practice of establishing new journals to meet the metric criteria has subsisted in countries whose current research evaluation systems were built on the ruins of the socialist research evaluation model, such as Poland (Kulczycki, Rozkosz, & Drabek, 2019) or Slovakia (Pisár & Šipikal, 2017). Similarly, in China, university journals play a significant role in scholarly communication (Liu, 2012), although they are, as in other countries, criticized for their local authorship and lack of disciplinary identity, because a single university journal can publish papers from all fields of science.

The number of scientific journals, and the number of publications and citations, is rapidly increasing. In the Ulrichsweb Global Serial Directory in April 2010, 69,262 active and academic/scholarly journals were indexed (Tenopir & King, 2014) whereas in May 2021, this number had grown to 140,410. Thus in just a decade, the number of active scholarly journals doubled. This dynamic growth reflects the pressures produced by publication-oriented academia and is driven by a lower technical entry threshold for new journals, emerging new subdisciplines of science, and pressure to increase the number of publications. Moreover, academic publishing is an important market which can generate significant profits for companies not directly connected with doing research. For instance, in 2010, Elsevier attained a 36% margin, which was higher than those that Apple, Google, and Amazon posted that year (Buranyi, 2017). The establishment of "Proceedings of University X" type journals, as of other journals, is spurred on by these pressures as well as business and technical possibilities. Nonetheless, one can identify certain periods of acceleration in the establishment of new journals by higher education institutions that are the result not only of a desire to communicate research or to make money but also of changing evaluation rules in a given country.

In Poland, following the Russian model, many "Proceedings of University X" journals have been established. Moreover, this practice did not end with the establishment of a journal for each university. Now, each department or faculty has one or more of its own journals. For instance, in 1965, the journal named *Zeszyty Naukowe Uniwersytetu Mikołaja Kopernika w Toruniu. Mathematical and Natural Sciences. Prace Stacja Limnologicznej w Iławie* [Proceedings of Nicolaus Copernicus University in Toruń. Mathematical and Natural Sciences. Papers from the Limnological Station in Ilawa] was founded to publish works by the employees of a single department. This journal is still continued under the name *Limnological Papers* because many journals are changing their names to avoid the term "Zeszyty Naukowe Uniwersytetu" [Proceedings of University] and to follow more "western" patterns of journal publishing, a trend that has also been observed in China (Liu, 2012).

One can note that the implementation of the new Polish Journal Ranking model and the counting of publications from all kinds of scholarly journals (local, domestic, or international) has accelerated this process (Krzeski et al., 2022). In the second decade of the twenty-first century, more than 100 scholarly peer-reviewed journals were founded each year in Poland. This trend slowed down slightly after 2019, as local journals were no longer as attractive in terms of university ratings. Nevertheless, in 2018, there were nearly 3,500 journals published in Poland. The overwhelming majority of publications in all science disciplines are published by Polish researchers, who number about 100,000. The sudden growth of Polish journals is clearly apparent: In 2012, there were 1,639 Polish journals in the Polish Journals Ranking, already in 2013, this had jumped to 1,807, while in 2015, it was 2,212 (Kulczycki et al., 2015; Kulczyck et al., 2019a; Rozkosz, 2017). In order for a journal to be included in the ranking, the only requirement was that they meet formal criteria, or more precisely, declare that they met them. Worthy of note is the fact that articles published in journals indexed in the ranking could be counted in the evaluation of universities within the national system. Therefore, for many universities and departments, establishing their own journals was the simplest way to "settle" with the government (as it had been since the first half of the nineteenth century in Russia). Since the government expects a specific number of publications from a department, the department will establish a journal, its staff will publish in it, and thus the department delivers the expected results. It is of no concern to universities that most of these journals are published late, sometimes with a delay of several years, for example, the required 2013 issues do not come out until 2016 (Drabek et al., 2017). The most important thing is that all the issues required are published before the time of the next evaluation exercise.

This is not just the practice of universities that could be classified as weak. My *alma mater*, considered the third university in Poland, publishes over 100 journals, and the Faculty of Philosophy alone, with less than fifty staff, publishes six journals. In most of these journals, publications by university staff make up the majority or a significant percentage of content. In the 2013 evaluation exercise, out of all the journals in which all Polish publications were published, the largest number (over 3,500) came from a single journal, *Prace Naukowe Uniwersytetu Ekonomicznego we Wrocławiu* [Research Papers of Wrocław University of Economics] which published ninety issues in just one year. Moreover, this practice does not only apply to journals but also to scholarly book publications. In Poland, every university has its own publishing house and what is more, many faculties or even departments at a university have their own publishing houses. The main purpose of these publishing houses is to publish scholarly books that can be referred to as achievements during academic promotions. In interviews conducted with directors of the largest university publishing houses in Poland (Kulczycki et al., 2019b), this practice was mentioned as the greatest obstacle for improving the quality of editorial and publishing practices. At the same time, interviewees stressed that it happens at the very explicit request of rectors and deans, who treat university publishing houses as "printing houses" that enable them to meet evaluation requirements during university evaluations or scientific promotions of individual researchers. The practice of establishing new university journals and transforming university presses into printing houses for their own staff in order to meet the requirements of the evaluation regime is a consequence of the demands put on higher education institutions. Deans and rectors are aware of the scientific potential of their staff, of how difficult achieving scientific mobility is in Poland, and how hard it is to attract the best scientists (Luczaj, 2020). Therefore, they choose methods for meeting the evaluation requirements that are not only feasible under such conditions (Krawczyk et al., 2021), but which, above all, will permit their institutions to maintain their status.

This is to be expected given that higher education and research institutions are organizations that are capable of completing resource-intensive tasks and achieving complex research goals while at the same time having to adapt to changing regulations and context. One of the key forms of adaption that is directly related to research evaluation systems is local use. This involves the adoption of national requirements and criteria at the local level, and the use of provisions designed for evaluating institutions in order to evaluate individual researchers. Through local use, over time institutions – in light of institutional isomorphism and the new institutionalism (DiMaggio & Powell, 1983; Dobija et al., 2019) – become more like other institutions in their sector, using similar tools to accumulate advantage and become highly ranked or evaluated (Dey et al., 1997). Local use in terms of research evaluation criteria has been observed in Australia, Denmark, Norway, Poland, and the UK, among others (Aagaard, 2015; Rowlands & Gale, 2019; Rowlands & Wright, 2019). Moreover, analyses of national systems reveal that national regulations are used in various local contexts and that this does not depend on science's position in a country. In other words, institutions from countries that are both at the forefront in terms of scientific productivity (such as Australia and Denmark) and from those that are lower ranked (such as the Czech Republic or Poland) play the evaluation game by using regulations from national evaluation regimes in local and institutional contexts. For instance, Hammarfelt and Åström (2015) shown that Swedish higher education institutions use allocation models based on bibliometric measurements, which is a product of imitation and a consequence of the fact that they operate under similar constraints. Cai (2010) investigated governance reform in the Chinese higher education sector and showed how global governance reform ideologies have spread in the Chinese higher education system, making different institutions similar to each other.

In some countries, however, similarities in terms of the regulations used in local contexts are remnants of the past. For instance, as suggested above, periodic evaluation processes of individual scholars in Polish higher education and research institutions emerged from a long socialist tradition. In an analysis of Soviet science policy, which significantly shaped the Polish academic sector, Cocka (1980) noted that evaluation tended to be formal and highly structured. Today, actual socialist government in Poland is a thing of the past. Nevertheless, when investigating research evaluation practices, one should take into account the possibility that some isomorphism such as local use, may result from a longstanding heritage of deeply embedded institutional practices and habits in the science and higher education sectors, and not from the adoption of similar solutions. In other words, institutions in post-socialist countries may tend to use national-level indicators in their intra-institutional, local assessments because, historically, they were encouraged to do so. And in fact, they do just that. Thus more than 70% of higher education institutions use national-level metrics ("Polish points") designed for institutional evaluation in their mandatory periodic evaluations of individual researchers (Kulczycki et al., 2021b). While the institutions are required to conduct such assessments, they have the autonomy to choose the assessment criteria. Local use in Polish institutions is not limited to any particular field and deans tend to use similar arguments to rationalise their decisions. They may therefore state that they believe that using national indicators in the assessment of employed researchers will lead to improvement of outcomes in national evaluations. This local use might be understood as a form of evaluation game. It is a direct response to the government mandating performance evaluations of individual researchers, as well as the national evaluation exercise. As DiMaggio and Powell (1983) showed, mimetic processes occur more frequently in organisations with large numbers of personnel and exert great pressure on the programs and services that they offer. This could explain why larger and highly ranked institutions used global indicators (instead of or in parallel with "Polish points") more often than do smaller ones.

Managers in academia who are seeking to maintain the position of their institutions while operating within the rules of the evaluation regime take advantage of all opportunities, including manipulating full-time equivalent employees. Since 1992, universities in the UK have been allowed to choose which staff they include in submissions for evaluation (Sayer, 2015). This propels universities to move researchers to teaching-only contracts ahead of the Research Excellence Framework exercise. *Times Higher Education* (Baker, 2019) showed that from 2015, a fifth of institutions increased the share of full-time teaching-only academics by at least 5 percentage points. It also asked three institutions what role the impending Research Excellence Framework played in these changes to teaching-only contracts. None of the institutions addressed this question directly which is unsurprising given that the manipulation of academics reported to evaluation was raised as an issue in the UK many years ago, and some institutions are considered "game players" (Jump, 2015). This also has other far-reaching implications. As Sayer (2015) indicates, university managers can synchronize decisions on resource allocation (for instance, the merging or closing of departments) to the cycles of the Research Excellence Framework. They may also, as Jump (2013) shows, employ researchers on mere 0.2 FTE (full-time equivalent) contracts, which is the lowest level of employment that qualifies academics to be included in the evaluation exercise. This is simply another form of the game that involves the offer of cash-in exchange for adding a second affiliation to papers by top-cited researchers. In a similar vein, Saudi universities rose in the global university rankings (Bhattacharjee, 2011) because they offered monetary incentives to a wide variety of academics from elite institutions.

When the Polish government decided in 2018 to change the unit of assessment in the research evaluation system from institutions (e.g., a university's department of physics) to disciplines (researchers classified according to their ongoing research as physicists regardless of whether they work in a physics or biology department), almost all Polish universities redesigned their organizational structures that had been in place for many decades. The only reason for this change, which was very clearly communicated by rectors, was to align departments in such a way as to fit most effectively into the evaluation system. Thus, for example, my *alma mater's* Faculty of Social Sciences was broken up into four new faculties: philosophy, psychology, anthropology and cultural studies, and sociology. In this way, through adjustments made by rectors and deans to metrics, the research evaluation system rapidly reshaped the Polish science and higher education sector landscape. This was similar to the decision by the communist authorities in the 1950s to establish the Polish Academy of Sciences, as described at the beginning of Chapter 2.

6.3 Adjusting Publication Patterns

In publication-oriented academia, which reduces research to papers and researchers to their authors, the impact of research evaluation systems is most conspicuous in publication patterns. These include changes in publication venues (local vs. international journals), types (scholarly books vs. journals), subject matter (trending topics vs. locally relevant topics), and language (English vs. local languages). If researchers are assessed by publication criteria built on specific international citation databases, or if articles are favored over books, one can anticipate that they will quickly work out how to adapt to the changing situation and play the game. The strength and mode of that adaptation depend, among other things, on a science system's evaluative power. For example, Schneider et al. (2016) show that the effects of evaluation regimes on the growth in the number of publications by scholars in Australia and Norway are different. In Norway, overall publication activity goes up and publication impact remains stable, whereas in Australia, publication activity also increases but the largest increase is in lower-impact journals which leads to a drop in Australia's overall citation impact.

Playing the evaluation game by changing publication strategies is especially evident (and dangerous) among the young scholars who are entering a highly metricized and economized academia. They have to adapt to the game and adjust their publication patterns just to access academia, get hired, and keep their jobs. Warren (2019) showed that new assistant professors in sociology have already published twice as much as their counterparts did in the early 1990s to get a job in top sociology departments or to get tenure. The race to publish as much as possible has therefore been on for many years. Müller and De Rijcke (2017) argue that in a highly metricized academia, researchers start "thinking with indicators." The outcome is that the planning of a research project, its social organization, and the determination of its endpoints are all adjusted to the metrics. Researchers act in this way in order to survive in academia and to hold on to their current position.

Research evaluation systems have created incentives for publishing in English and for preferring articles as the dominant type of publication. This pattern is widespread in all countries where such systems are in place. For instance, in the Czech Republic, publication patterns, and publication languages may be directly influenced by the national evaluation system, which influences 100% of the core funding of research from the state budget. The Czech evaluation system encourages the production of articles published in Scopus and WoS journals and proceedings (Petr et al., 2021). Thus, an article published in a top international journal obtains several times more points than an article published in a journal that concerns "national" fields (e.g., law), focuses on local topics, and is written in Czech (Good et al., 2015). An analogous situation may be observed in Poland and Slovakia as well as in Denmark and Belgium (Aagaard, 2018; Engels & Guns, 2018).

Changes in publication patterns are driven by global processes of metricization and economization, and the competitiveness of the scientific market. However, they may be accelerated by national or institutional contexts, as already documented in numerous studies (Aagaard, 2018; Aagaard & Schneider, 2017; Butler, 2003a; Engels et al., 2018; Kulczycki et al., 2018; Sivertsen, 2014; Thompson, 2002). One of the most conspicuous signs of these changes are the many voices raised against scholarly book publications because they are hardly taken into account in academic evaluation systems. In particular, book chapters and festschrifts (as discussed in Chapter 1) are presented as types of publication to avoid. Thus some research-ers predict the disappearance of scholarly monographs, which they have attributed explicitly to research evaluation regimes (Williams et al., 2018). Nonetheless, in some humanities disciplines and parts of the social sciences, publishing a schol-arly monograph is a requirement (or a strong expectation) for obtaining tenure. For instance, in History, a monograph is considered a test of competency and of pres-tige, and a necessity in order for obtaining tenure in the United States (Townsend, 2003). For Europe, Engels et al. (2018) show that despite those voices arguing that monographs are on the way out, scholarly books are not dying out in the humanities and social sciences. There is, all the same, no doubt that even in the humanities, journal articles are becoming the key channel of scholarly communication.

The increase in the importance of articles appears rapid and significant when one is drawing on global databases such as WoS and Scopus. However, a study based on comprehensive national data on all publications by researchers from eight European countries (including all types of scholarly book publications) shows that publication patterns are stable (Kulczycki et al., 2018). In addition to a large increase in the number of articles in international databases from particular countries or institutions, there are also increases in international collaborations, the average number of co-authors, and the number of scholarly book publications (Olechnicka et al., 2019). Simply put, researchers are publishing more and more of every type of publication.

Researchers are compelled to publish as much as possible in English, even in the humanities and social sciences. The logic of the internationalization of science, together with research evaluation systems, has resulted in English being perceived as the key language of scholarly communication. This poses substantial challenges to researchers outside the Anglo-Saxon world. In this regard, I agree with Sivertsen (2018a) who argues that the use of local languages is necessary in scholarship in order to foster engagement with stakeholders and the publication. However, where evaluation regimes influence practices and modify research agendas, researchers may choose to move away from locally relevant research toward decontextualized approaches that are of interest to English language audiences. This can cause great harm to research and scholarly communication because, as López Piñeiro and Hicks (2015) demonstrate, the interests of international and domestic readers differ. Moreover, as Chavarro et al. (2017) contend, the goals of non-English versus mainstream English journals are also distinct. The former give researchers the opportunity to be initiated into publication and to publish on topics that are not well covered in mainstream channels.

Research evaluation regimes oblige researchers to publish in English, and the assessment of individual researchers is often based on international indexes, and on a tally of the number or share of publications in English. This distorts the picture of actual publication practices, especially of researchers from the social sciences and humanities in the Global East. Thus only less than 15% of publications by Polish researchers in the humanities and social sciences are indexed in the WoS, 16% of Slovenian researchers' publications, and 50% in the case of Danish researchers (Kulczycki et al., 2018). An overview of publication patterns that does not take into account 85% of the data cannot be correct. The European Network for Research Evaluation in the Social Sciences and the Humanities analyzed bibliographic data from national current research information systems from eight European countries. Through this research, we found that in 2014, 68% of all publications in the humanities and social sciences in Finland were published in English, in Denmark the figure was 63%, and in Norway, nearly 62%. At the other extreme were Polish scholars, only 17% of whose publications were in English, compared to 25% in Slovakia and 26% in the Czech Republic. However, even this picture, based on national systems, was not complete. Our initial analysis reproduced an approach that reduces scholarly communication and its evaluation to the counting of individual publications, adding up the total number of publications and the percentage of publications of a given type or in a particular language. However, it is individual researchers who publish papers and who are at academia's core, not the publications themselves. Thus when, after two years, we looked at researchers' publishing practices in relation to articles, we found that the poles were reversed (Kulczycki et al., 2020): Nearly 70% of Slovenian researchers in the humanities and social sciences published their articles in at least two languages (10% in three languages) during the three-year period analyzed, 57% of Polish researchers published in at least two languages, with only 39% of Norwegian and 43% of Danish researchers doing the same. English has always been the second language after the local language. Thus the above shift in perspective reveals a completely different picture, which should change policy because it points to a different pattern: Researchers from CEE countries are internationalized (publishing in at least two languages) and locally oriented. And yet, science policy that is focused on publication in English has a significant impact, pushing researchers not only to try to publish mostly or only in international journals but also to redefine their research topics.

The conditions that constitute the evaluation game have transformed the way in which one publishes. Thus not only are the what, where, and how many of publication have changed but also, the question of whether what one does is considered science at all is now open to debate. The latter has occurred because evaluative power creates homogeneity, that is a standardized and uniform set of criteria for research assessment which reduces epistemic pluralism (Bonaccorsi, 2018b; Krzeski et al., 2022). Pardo Guerra (2020) demonstrates that the Research Excellence Framework in the UK led to reductions in thematic diversity within sociology and a shift in the thematic composition of departments across all disciplines. Wieczorek and Schubert (2020) show that sociologists adapt and play the evaluation game by focusing on mainstream topics and apply strategies of publication management which result in low research diversity. In this way, they try to become "REFable" researchers, that is scholars who produce outputs that are highly rated in the evaluation system and thus are employable. This shows how researchers not only think with indicators but also constitute their academic identities through them. Moreover, one can find evidence of an even stronger REF impact in economics. Within the discipline, heterodox economics is being severely marginalized, to the point of being pushed out of the discipline entirely, which fact, as Stockhammer et al. (2021) argue, endangers pluralism in economics research. This development is due to the fact that the journals that publish top-rated REF articles in economics do not publish heterodox economics research.

The consequences of the playing of the evaluation game on epistemic pluralism are becoming ever more visible. In a piece on the Mexican ecological research community, Neff (2018) examines the structural influences exerted by research evaluation regimes and monetary incentives. Thus the Mexican research evaluation system impacts the selection of research topics and field sites and drives researchers to publish in English language journals. Neff refers to the case of a researcher who studies the evolution of agricultural crops and collaborates with communities of small-holder farms that grow native varieties. This researcher highlights that his work is not publishable in the impact-factor journals that are highly incentivized by the Mexican system, because those journals publish research that leads to the commercialization of crops. Thus despite its quality, his research is not in line with the current priorities of those journals that are treated as the benchmarks for quality in Mexico. Thus, "he finds himself balancing the work that he values based upon his commitment to the farming community with that which is publishable in high-IF journals because the base salaries are insufficient without the SNI bonuses" (Neff, 2018, p. 198). Research evaluation regimes put researchers in a position in which they must try to forge a balance not only between institutional and disciplinary loyalty but also between local and international audiences. Each evaluation technology that incentivizes publishing mostly in English leads brings about a narrowing of local research topics that are crucial to the local community but that may be of no interest to readers of international journals. The situation is such that even including the name of a country or region in the title or abstract of an article can reduce its readership and citations (Abramo et al., 2016; Jacques & Sebire, 2010).

In the course of their work, researchers are confronted with the challenge of fulfilling contradictory expectations. On the one hand, they must work to solve local problems in order to produce a social impact, while on the other, they are pushed to publish papers in internationally recognized journals in order to prove their research excellence. One of the more commonly adopted strategies in response to this dilemma is to decontextualize research problems, that is, to address more universal issues that are not embedded in a local context. In addition, researchers who are discussing issues related to their country of origin also become more likely to cite scholars from the UK or United States, even when scholars from their own country have more relevant work on the topic (Rowlands & Wright, 2020). However, this pattern can also be altered by an evaluation regime. For example, Baccini et al. (2019) show that the implementation of a national evaluation system for individual scientists in Italy in 2011 caused Italian scientists to self-cite themselves more often and to form "citation clubs" of Italian researchers who quote each other.

* * *

The three issues discussed in this chapter have highlighted the different aspects of changes in researchers' publication patterns that occur as a result of playing the evaluation game. They also show that researchers adapt to evaluation rules because engaging in redefined practices is more profitable than reproducing old patterns of behavior. The actions that researchers initially take because of external evaluative pressure or incentives, over time take on a life of their own. Thus researchers internalize the new rules of the game and begin not only to act according to them but also to think through them.

The playing of the evaluation game is made possible by the pressure on all researchers and managers of scientific institutions that fosters a social norm in academia in which playing the game is valued as something positive. The game gets played because players produce the right outputs for their institutions while maintaining their position in the system. The stronger the evaluative pressure and institutional loyalty, the easier it is to create a socially shared belief that it is appropriate to publish in venues that are beneficial to one's institution, even when this conflicts with the values shared within the scientific field.

When researchers commit fraud or falsify data, they face legal consequences. When they game the system by salami publishing or being credited with authorship of articles in which they had little contribution, their actions are condemned as violating scientific ethos. By contrast, most of the time, players in the evaluation game do not get condemned. Usually they are completely invisible in the global academic field and in the local context, conforming to the rules of evaluation understood as a necessity arising from external pressure. This further means that the cultivation of institutional loyalty at the expense of disciplinary loyalty does not get penalized. Additionally, there is no system of punishment in place that recognizes publishing in questionable journals (Kulczycki & Rotnicka, 2022). Neither is the abandonment of the study of issues important to local contexts viewed as transgressing legal or professional boundaries. Thus as I have shown in this chapter, the playing of the evaluation game has repercussions far beyond the mere dynamics of the evaluation process itself. In addition, it is also becoming a fundamental way of doing and communicating science that is the product of the large-scale metricization and economization of science.

As I have shown, the players in the evaluation game are not only researchers but also deans, rectors, and directors of university presses. There are no penalties for players because the playing of the game is expedient for key players. Above, I discussed the case of a researcher who, in the Polish evaluation system, reported the publication of fifty-three monographs in a four-year period. These monographs were released by the publishing house of the university which employed this researcher and which, moreover, paid for reviews of the monographs (the total length of each of these reviews was two sentences, which were almost identical for each review). Moreover, these reviews were performed and the books published only a few weeks before the end of the reporting period; thus the publication of these books reinforced the institution's evaluation score. One might hypothesize that this researcher's concern for improving the institution's evaluation score stemmed from the fact that he also happened to be its director.

Conclusions

In this story about the forces remaking research, I surveyed global phenomena and local responses to them, observing managers, researchers, publishers, and policy makers at work. Charting the transformation of the entire science and higher education sectors, I highlighted the influence of new metrics such as the Journal Impact Factor, or the shift to a grant model for funding researchers' work. The outcomes that are pledged as fruits of an evaluation regime, notably greater accountability, transparency, and productivity, have indeed materialized. At the same time, however, these regimes have distorted not only how researchers communicate the results of their research but, more importantly, what researchers consider science.

I have shown that with the proliferation of technologies of evaluation power and the expansion of power relations between states, institutions, and researchers, researchers' everyday work practices around publishing their findings are being altered. This manifests itself in the fact that researchers are beginning to align their communication behaviors with evaluation guidelines. These changes are sometimes seen as misappropriations of the scientific ethos, or labeled outright as unethical behavior in pursuit of extra benefits. To capture this new dimension of the everyday work of researchers in a more heavily metricized academia, in Chapter 1 I introduced the key concept of the *evaluation game,* which organizes the book's narrative. I defined the evaluation game as a practice of doing science and managing academia in a transformed context that is shaped by reactions to and resistance against evaluative power. When new rules and metrics are introduced, researchers and managers devise various strategies for following these rules at the lowest possible cost. These strategies, as forms of adaptation, are reactions to new rules and metrics.

In the following chapter, I argued that research evaluation systems are the result of an altered science and higher education landscape that is being transformed primarily by metricization and economization. Economization promotes the idea that science's economic inputs and products such as publications should be utilized for bolstering the economy, while metricization reduces every aspect of science and social life to metrics that serve to measure science's products. I used Chapter 3 to illustrate how the picture that is usually presented in studies on research evaluation is incomplete because it mostly ignores the legacy of Imperial Russia, the Soviet Union, and the countries under its influence. Thus I showed that the publish or perish research culture did not only emerge with the birth of neoliberal market logic in higher education. Another of its sources can be traced back to the Russian drive to bureaucratize universities and the subsequent use of these schemes in the Soviet Academy of Sciences and other national academies established on this model. Building from this foundation, I reconstructed two models of research evaluation system, the capitalist and the socialist. I further described how the latter should be regarded as a national research evaluation system that was established several decades before the emergence of performance-based research funding in the UK during Margaret Thatcher's time. In Chapter 4, I put the approach to evaluative power and its technologies outlined at the start of the book to work. Thus I presented several national research evaluation systems (including those in Australia, China, Poland, and Russia) and highlighted the ways in which metricization and economization shape academia.

The use of metrics and the focus on increasing productivity impacts the weakest persons in academia most severely, notably young researchers and researchers from peripheral countries. As a consequence, a logical response to this expansion of evaluative power might take one of two tracks. It might seek either to (1) improve the metrics themselves and the way they are used or (2) abandon the use of metrics in science, knocking these oppressive tools out of the hands of senior managers and policy makers. Attempts to implement both kinds of response are indeed being made. The trouble, however, is that in this particular historical moment, both approaches are doomed to failure. This is because the causes of the metricization and economization of science are neither metrics' imperfection nor their use per se. Rather, the problem lies in individualized thinking about science and the focus on the accumulation of economically conceived value by institutions in the science and higher education sectors. Therefore, this book's ultimate assessment cannot be a recommendation for the "responsible use of metrics," nor a call to abandon the use of any metric in academia. A third answer is needed. The chief contribution this book seeks to make is its call for a rejection of these two inadequate responses, and its proposal that we set out on a process that offers up hope of a third way in academia. Before I explain why these two paths will not free us from the pervasive regime of imperfect metrics, and how we can move in the direction indicated by a third way, I summarize what I learned while working on this book, and what the reader might take from it.

In starting to write this book, I set myself the goal of offering an alternative position to the discourses on the use of metrics and the measuring of science that are being produced by research evaluation scholars. This desire to present an alternative perspective grew in me with each passing year as I worked on evaluation systems and their effects. I frequently discussed transformations of the science and higher education sectors in international groups of evaluation scholars or policy makers. Every time I did so, I noticed that my colleagues assumed that the current publication pressures in academia, including toward greater accountability, and the increased institutional demands on researchers to produce more publications were the result of the neoliberal transformations of the last three or four decades. At the time, I realized that this assumption of a homogeneous source for the growth of evaluative power causes many researchers to view the effects of research evaluation systems, and how researchers respond to being evaluated, in a uniform way. In other words, wherever there are research evaluation systems based on publication counts, the Journal Impact Factor or other bibliometric indicators, researchers will react in the same way. And yet, in my everyday work in a medium-sized European country, and through my governmental advisory role on the construction of a national research evaluation system, I observed that Polish researchers reacted and, importantly, adapted to the evaluation regime differently than did researchers in other countries. This sparked my interest in the specific way in which Polish researchers respond to the evaluative power produced by both global institutions and the national evaluation regime. Thus I was led to thinking about the playing of the evaluation game as a response (different from gaming) to the changing conditions of scientific work in an academia that is subject to intense metricization and economization.

In searching for the origins of this different response, which is evidenced through the enhanced institutional loyalty of researchers at the expense of loyalty to scientific discipline, I turned to the modernizing potential of measures and metrics used in Russia and then the Soviet Union. This allowed me to show that metrics for monitoring and evaluation in academia were implemented in Russia more than a century and a half before the emergence of neoliberal logic in the science and higher education sectors. Thus, starting from my curiosity about Polish researchers' distinctive reactions to the apparently identical metrics that are used whether in Australia or Spain, I arrived at the origins of the first national system of research evaluation in the Soviet Union. This was an *ex ante* evaluation system that was focused primarily on the question of whether the research undertaken would be consistent with the values and policy goals of the state.

From the outset, one of my aims has been to show how differently post-socialist versus non–post-socialist realities shape global and national evaluation technologies and responses to them. I also drew attention to the fact that players game the system in the same way everywhere in order to obtain additional bonuses. However, the evaluation game takes place in diverse geopolitical contexts that result in its playing being different in Poland and Australia. When I began to write the book, it seemed to me that researchers in Poland and Australia play the evaluation game differently. Now, at the end of this journey, I still believe this to be the case. Yet as I have tried to show, what differs most in the reactions of researchers, for instance, in Poland and Australia or the UK, is their attitude to metrics and level of trust in experts, rather than the way the game is played. Thus it is not so much a question of variation in terms of how players play the game, but rather, of how historical context has shaped the evaluation game, and how players perceive evaluation systems.

This faith in metrics in a country like Poland is, in my opinion, evident among all kinds of players, from policy makers who prefer to cede power to metrics rather than to experts, to institutions that favor clear evaluation guidelines and information about how many points they will receive in exchange for the minimum amount of scholarly activity. In addition, one can observe it among researchers who believe that metrics will protect them from the partisan interests of their opponents in the science system, and finally, publishers and editors who prefer to conform to journal evaluation rules rather than strengthen a journal on the basis of a particular scientific discipline's values.

Having already defined and described the key pieces of the puzzle, notably the evaluation game, evaluative power, and its technologies as used in different countries, and having emphasized that a technology's context of implementation can make a significant difference, I have shown that there is a broad spectrum of responses to power and technologies. In Chapter 5, I showed that the term "gaming" is inadequate to the task of capturing how researchers change their scholarly communication practices under the influence of publication pressures and the metrics used to evaluate their work. This is because researchers who play the evaluation game do so not only to increase their profits but also to keep their existing position; this is their primary motivation. Further, in this chapter, I described the primary motivations of all the key players, that is, researchers, institutions, publishers, and policy makers. In Chapter 6, I used the theoretical tools developed thus far to analyze three areas in which researchers play the evaluation game – publishing in predatory journals, following the metrics, and changing publication patterns. In doing so, I described how the evaluation game is not an all-encompassing phenomenon but is rather constrained, among other things, by the goal of preserving the status quo by pursuing the demands of the evaluation regime. Therefore, this concept can be useful for identifying and understanding a different range of scholarly communication practices than those encompassed by fraud and gaming.

The path traced by the book casts light on current developments in academia, shaped as they are by the logics of economization and metricization. As noted, one can name two broad approaches to the problems/difficulties generated by metrics. Let us now examine the validity and feasibility of these two options through the perspectives offered by the book.

Two Logical Responses to the Proliferation of Metrics

Many scholars believe that the metricization and economization that are currently dominant are based on sound and useful foundations and that only certain elements of the overall structure they are a feature of need improvement. They argue, therefore, that all that is needed is minor tinkering at the peripheries of the science system. Thus they might prescribe a little more transparency and the highlighting of the flaws and limitations of bibliometric indicators for one area. In another, they might promote mild regulation to limit the power of global rankings, while in yet another, they might recommend a minor change in the funding system for researchers so that competition for grants does not turn bloodthirsty. However, as I have sought to demonstrate, it is the very assumptions that underly the use of metrics in science that are deeply problematic. These assumptions promote the counting of publications and citations and an approach to the results of research communication that treats them as products to be weighed and measured and likely to generate profits. Neither of these two reasonable responses to the proliferation of metrics will suffice if they do not involve a complete rethinking of science to view it as a collective enterprise that is geared toward meeting the needs of the society from which researchers originate, and which in turn funds that science.

Response 1: Improving Metrics and How They Are Used

When I started studying national research evaluation systems and the effects they produce more than a decade ago, the remedy for the metricization and economization of scholarly communication (the main drivers of evaluation power) that was promoted was better metrics, that is, so-called altmetrics based on social media (Priem et al., 2010), and a change in the publication model (Suber, 2012) from subscription-based (readers pay for access to publications) to open (authors or the institutions funding their work pay for access, which is free to readers). For many years now, the Journal Impact Factor has been criticized for its use as a key bibliometric indicator for evaluating both journals (as intended) and researchers or universities (which it was not designed to do). Within bibliometrics and scientometrics, new and more sophisticated metrics have been proposed for many years to describe and evaluate the whole spectrum of scientific activity: from interdisciplinary research to the evaluation of research groups and the linguistic diversity of journal articles. Responding to the weaknesses of existing bibliometric indicators and the negative consequences produced by them, researchers started to call for improvements to the indicators and to propose new ones, such as the h-index proposed in 2005 by Hirsch (2005). Today, we already have hundreds if not thousands of metrics in science and they are used and promoted in increasing numbers by global giants such as Clarivate Analytics and Elsevier. Moreover, thanks to tools like InCites or SciVal provided by these companies, any dean or rector can search for their employees in bibliometric databases and assess their "usefulness" and "excellence" in scientometric terms. Thus, a key element of the first response to the proliferation of metrics is the creation of more metrics that do not reduce scholarly activity to publications alone but also take into account international collaborations, the level of sophistication of open science practices, or the economic or social impact of the research conducted. Along with this call for more metrics comes a second element of this response: a call for the responsible use of metrics.

In 2013, the San Francisco Declaration on Research Assessment (DORA, sfdora .org) was published. It calls for an end to the use of the Journal Impact Factor in research assessment. I frequently encounter critics of the use of any metric who claim that DORA calls for the abandonment of all metrics in research evaluation. This is not at all the case, however. DORA calls for abandoning the use of the Journal Impact Factor in areas for which it was not intended, that is, for evaluating scientists or institutions. There is no question of rejecting other metrics. Moreover, DORA explicitly recommends that organizations that produce metrics should behave transparently by releasing the data and methods used to calculate all metrics. In other words, DORA proclaims that metrics are okay if you use them responsibly. In 2015, *Nature* published the "Leiden Manifesto" (Hicks et al., 2015). Its authors outlined an approach for improving research evaluation and implementing it more responsibly. The last of their ten recommendations refers directly to metrics, which, according to the authors, should be scrutinized and updated regularly. The same year also saw the publication of *The Metric Tide* report (Wilsdon et al., 2015), in which – as a response to the DORA and "Leiden Manifesto" recommendations – the concept of "responsible metrics" is proposed and defined as "a way of framing appropriate uses of quantitative indicators in the governance, management and assessment of research" (2015, p. 134). In other words, in the eyes of many researchers, the cure for metricization and economization may be new metrics, disclosure of their modalities, and responsible use.

I am certain that this rational, balanced, and evolutionary response to the challenge of transforming science under the influence of metrics is reaching the ears of many scientists, policy makers, and senior officials in academia. I make no secret of the fact that this is the path I followed for years and indeed contributed to by proposing "more accountable" solutions for various research evaluation systems. I have persisted on this track and will probably continue to do so for some time, as it offers some hope for some change and, importantly, unites critics of metrics in positive and constructive action: new, better metrics mean new and better outcomes. Working on this book has enabled me to gain a multidimensional perspective on the use of metrics. At the same time, it has also made it extremely clear to me that any effort by academia to improve its situation by collaborating with the commercial companies that control scholarly communication will end in failure for academia. Commercial entities will eventually absorb all of academia's collective efforts to build responsible metrics and privatize them for their own financial gain. This is precisely what happened to the idea of open science which promoted openness not only for scholarly communication but also for all scientific practices, so that they would be universally accessible and subsequently usable by researchers and the general public. I was, from the very beginning, an ardent supporter of open science. However, I noticed that open science gradually got reduced to open access to publications. This continued to such an extent that eventually, having started out as a very noble and necessary idea, open access became a tool used by the largest publishers to multiply their profits.

What is worse, open access to publication has become a luxury for researchers in central countries who can afford to pay Article Processing Charges. Despite the existence of various discounts, young researchers from peripheral countries cannot afford to pay these fees, which leaves them to publish either in subscription journals or in open-access journals outside the mainstream of scholarly communication. Thus this publication access model divides journals into two groups in the same way as the Journal Impact Factor, which has been criticized for decades: those that had JIF and those that did not, but wanted to. Thus, I believe that while calling for improving metrics, exposing their limitations, and demonstrating responsible ways to use them is necessary and useful in the short term, it is entirely inadequate in the long term. While we should improve these metrics, we must also prepare ourselves for the long road ahead that addresses another front. That front cannot, however, be the one indicated by the second response to which I now turn.

Response 2: Stop Using Metrics

Since the root of the vast majority of problems in academia is its metricization and economization, we should therefore rid science governance of metrics. Such a demand is today growing louder in many circles (Hallonsten, 2021) and is presented as a key remedy for healing an ailing academia. If all metrics eventually become the goal rather than the measure, then the use of metrics in academia should cease. At the heart of this call is the assumption that academia will do better without metrics and that we will return to the original state of no citation and publication counts. Only these assumptions are wrong: There has never been a time in academia in which universities or research institutes were uncontrolled. Of course, the number of citations or the number of publications were not always the tools used for this purpose (although, as I have shown, in Russia this started already two centuries ago). However, it has always been possible to identify certain metrics that the institutions funding research used to verify that their interests were being adequately catered to. For example, this could involve the education of specialists with skills and knowledge appropriate to state administration, or of engineers whose knowledge and experience served the country's development.

Thus just as there has never been an ivory tower for researchers to work in that has no contact with the public (Shapin, 2012), there has never been a state of "no metrics" in academia. This is due to the fact that since the very beginning, metricization was linked with modernization forces. Therefore, it is not a matter of looking for a moment in the development of the science and higher education sectors where citations, the number of publications, or number of grants were not used to hold the sector accountable: One will of course find such a historical moment. Yet for those who call for the abandonment of metrics, the aspiration is for a full autonomy, in which no forces outside the sector have decision-making power within it. And yet this has never been the case, because, since its inception, academia has been – like any other institution – bound by multiple power relations.

Often, calls for the rejection of metrics simply take the form of total negation, that is, let's reject metrics, and let's not bother to indicate how to manage the sector differently. However, it cannot be said that all such suggestions are devoid of a constructive element in terms of pointing out "what next" and "what" metrics would be replaced with. Most often, metrics are replaced by expert evaluation or peer review, which is portrayed as the "natural" form of research evaluation, and which can be used as a means of managing the whole sector. However peer review as we now know it (mainly through the evaluation of manuscripts submitted to journals or assessments of grant applications) only became widespread in the second half of the last century (Chapelle, 2014; Moxham & Fyfe, 2018). Moreover, for many researchers, it is metrics that became the remedy for many of the imperfections of peer review with its various types of bias, as I highlighted when describing the research evaluation systems in Russia or Poland.

I believe that one cannot realistically move beyond metrics because they are not a defining feature of academia, but of capitalism itself. Metrics, metrics-based evaluation, and rankings are used to justify decisions in an economy of scarcity: of funding, top researchers, aspiring and promising students. Thus, metrics are used to justify the taking of resources away from one institution or scholar to give to another. Therefore, any reform of metrics in order to promote their more responsible use, is, by necessity, undertaken within the framework set by metricized capitalism.

Calling for the abandonment of metrics based on citations and publications will change nothing. New metrics will be found that are based, for example, on the degree of openness of the results of scientific papers, and these will have to transcend the limitations of previous metrics. This entire process will continue to unfold, because its driving force is the desire to accumulate advantage in terms of available resources and prestige, which can be expressed precisely through metrics. Accumulation is achieved by inducing and enhancing competitiveness: In academia today, both institutions and researchers are seen as the sector's "discrete atoms," which can provide further resources if pressed hard enough. As a result, academia is seen through the prism of publications rather than scientific discoveries and publications viewed as the product of their authors rather than the researchers or groups behind them. It is not so much a question of metricization and economization making the individual (institution or scientist) central to today's thinking about science, but rather, that the individualization of thinking about science has opened the door to the metricization and economization of these sectors. Global thinking about science is based on the assumption that science is a collection of individuals who alone can expand the economic capital and prestige resources of an institution. Thus, such resources can be amassed if each individual (whether a single researcher or an institution) can be isolated and then weighed, measured, and evaluated. Alternatively, metrics may be altered (the solution indicated by the first response) but will still be based on measuring individuals as discrete units and comparing them to each other. Individuals will be judged by how they contribute to the accumulation of the economic or social capital of an institution, region, or country.

At the beginning of this book, I wrote that academia is constituted by people working together in institutions. It is academia as a collective entity that is subject to various processes and logics such as metricization. However, the focus on expanding advantage, and one institution taking over the resources of another makes us see "individuals" in academia instead of "people working together." Just as in scientific discovery we stand on the shoulders of giants or past scholars, so too in academia today, learning takes place in networks of relationships and collaborations that cannot easily be ranked because these networks are not discrete units. The discontinuous nature of these units (e.g. a scholar represents one discipline and may not represent any other discipline) makes grading and measuring flawed but simple, and thus understandable for many. We can see this in the difficulties posed by any attempt to measure and evaluate interdisciplinary research (Leydesdorff et al., 2019). It is far easier to monitor, supervise, and evaluate partnerships that are atomized rather than networked. It is also more difficult to instigate resistance to evaluative authority, as resistance is initiated mainly by individuals with the highest standing in academia who have a stable position. For resistance to be effective, it must be based on collective rather than individual action. However, the hierarchical nature of academia makes it very unlikely that universal and collective resistance, notably the rejection of the use of current metrics, will occur today.

Toward a Third Response

The economization of science is still the strongest driver of the global science and higher education systems today. At the same time, together with metricization, it is the most powerful transformative factor for these sectors. And yet it is important to remember that the power of metrics and economization will not endure if we, as societies and citizens of countries, place limits on it. Economization and metricization are logics of social processes, and as such, they require control and regulation. Like any logic, they can either improve a state of affairs or make it worse. For example, metricization can foster transparency in university funding, but at the same time, it can generate pressures on researchers that ultimately lead to tragic consequences in the form of suicide. I firmly believe that academia and the scholarly communication system can function more effectively if they can be shaped by a change in the attitudes of researchers, managers, publishers, and policy makers.

We will not escape metrics, but we must put an end to perceptions of science as reducible to the publications of individual researchers working at individual institutions. It is also not possible to return to a situation in which there are no metrics and measures in the management and organization of science, because such a situation simply never existed. As I have showed, metrics have always been used either as tools of modernization or of monitoring and oversight. What we can and should do therefore is deindividualize thinking about science and focus on the needs of our communities and societies. Let the use of metrics be guided by our goals, values, and beliefs, and let us not change our values and beliefs to achieve the goals set by metrics.

I want to make clear that metrics are useful. They are useful because they universalize the ability to measure and evaluate research across institutions and countries, thus overcoming particularisms and local contexts. However, this feature of metrics also makes it very easy for them to be appropriated by economic actors and incorporated into global monitoring and control tools. Science is a global endeavor, which is why we need global metrics. However, we cannot tolerate a situation in which public institutions and states produce publications and metadata about those publications, and that private companies aggregate that information, produce metrics, and sell them. Nor is it acceptable, more significantly, that based on these metrics, countries, institutions, and researchers are measured, ranked, and evaluated. Not only do public institutions and states not really know what these metrics show (global companies, while proclaiming the transparency of metrics, at the same time rely on a lot of understatement), but researchers and evaluators have not been involved in the construction of these metrics.

In what follows, I sketch seven principles that I think should be kept in mind when building not only a new system of scholarly communication but, more importantly, an academia that is not driven by capitalist metrics. When I began investigating research evaluation regimes, and also while working on this book, I hoped to identify some gap in the system that had been overlooked by others. I deluded myself into thinking that maybe there was a solution that could easily remedy the situation in academia today by fixing the technology of evaluative power. And yet, because metrics are not the problem, there is no such loophole. The solution is both in essence simple and at the same time a great challenge to implement because it violates the existing interests of many stakeholder groups.

While researching this book and in conversations with a large number of researchers, policy makers, and evaluation system developers, I have come to understand that efforts to improve academia should not be channeled into fixing or discarding technologies of evaluative power, which is what the two responses sketched above indicate. Rather, they should be focused on reformulating power relations. By these, I mean the relationships between the state, institutions, and researchers, and between all of these and global economic actors. The latter have the capacity to homogenize particularistic solutions that are designed to serve local environments and turn them into universal tools for capital accumulation. In doing so, they also shift the costs (i.e., the unwanted effects of using metrics or bibliometric indicators) onto states, research institutions, and researchers.

Let us start here: *we should foster an academia that brings out the best in researchers and managers, not the worst.* Concern for self-interest pushes researchers to play the evaluation game, as perceptions of the results of their work are filtered through the prism of economic values. Academia should enable scientists to pursue their own self-interest, but it should also encourage striving toward other goals that are to the benefit of society. We should therefore reward not high productivity but, above all, solidarity and integrity. We should, furthermore, also reward contributions to building trust in science and cooperation between researchers, and between academia and society. Reform of academia must be reoriented toward long-term thinking. Evaluations conducted every one, two, or even four years cannot tell us anything meaningful about the ramifications of research over time. Activities that benefit the common good should be defined within a long-term perspective (of at least a decade) and should be supported by government grants.

Second: *There should be a dramatic increase in stable funding for science through block grants.* Metrics are such a common feature of science and scholarly communication because academia operates under the framework of a capitalist economy of scarcity. The negative consequences of using metrics arise because scarce funds and resources must be distributed among a growing number of researchers and institutions. The first step on the third way, then, is to provide funding on an unprecedented scale. This does not mean an increase of 1% or 2% of a country's budget. It is necessary to go much further, as did countries like the Soviet Union and the United States, which sought to achieve advances through science and research in the middle of the last century, and as China is doing today. This increase in funding needs, however, to be rolled out everywhere across the globe. Its urgency is only underlined by the climate crisis and the havoc caused by the COVID-19 pandemic, both of which make it patently clear that it is science that allows us to deal with the great challenges. Yet we cannot act in an evolutionary manner, nor on the assumption that researchers will first demonstrate the worth of their work and account for the resources they use, and that governments may only then provide funding. A radical, even revolutionary, increase in science funding is necessary, not only in order to combat the negative consequences of metricization, but above all, to transform and ameliorate the quality of life of all people.

Third: *academic institutions should guarantee stable employment conditions and good salaries, including for early career researchers.* One cannot expect full dedication to research, and the development of scientific passion and scientific discovery if researchers are constantly thinking about the fact that their contract is about to run out. Academia's current competitive nature, with its grant employment and short-term contracts, does not create the conditions for normal work. And research is a job. Of course, researchers are very often motivated by a desire to fix the world, to improve people's quality of life or to understand the principles that operate in the natural and social worlds. They also expect recognition, aspire to work in prestigious institutions, and publish in the best journals. However, prestige of this kind does not pay the bills or feed the family. And further, it does not always trickle down to all who deserve it. I am convinced that any revolution or evolution in academia and scholarly communication must begin with a stabilization of the material working conditions of researchers and of the resources they have access to in their work. Creating secure workplaces does not imply that researchers cannot be managed, that nothing can be demanded of them, and that they cannot be fired when they do not perform as expected. On the contrary, it means that you provide researchers with safe and regular working conditions, clearly communicate expectations, and set rules which, after a few years, the funding institution will be able to check as successful. In creating decent working conditions, expectations and evaluation methods must be clearly communicated from the beginning, with responsibility for doing so in the hands of those who have evaluation power, not those who are being evaluated.

Fourth: *Researchers should be fully involved in defining evaluation criteria and producing metrics if the evaluation is to be based wholly or partly on them.* This is a necessary condition for building confidence in evaluation. Moreover, it would enable the design of evaluation rules that take into account scientific practices specific to different scientific fields and geopolitical contexts. No evaluation procedure can be trusted unless it is built in consultation with those who will be evaluated. It is not a matter of those being evaluated choosing the evaluation criteria themselves and defining the requirements they need to meet in a given time frame. This would only produce a sham evaluation. The point is instead that from the very beginning, those who will be doing the evaluating should state which areas of the researcher or institution's activity they are interested in, and make proposals for specific procedures. In the dialogue that is produced by such an approach, detailed solutions, metrics, and methods of expert judgment emerge, as do, critically, the areas to be ultimately addressed in the evaluation, and the consequences of the evaluation itself. On more than one occasion, I have found that, from the perspective of evaluation system developers, it is sometimes impossible to see what really matters to researchers in a given discipline and to distinguish between that which is presented as a key activity and that which is presented as merely an incidental outcome. Therefore, facilitating such dialogue from the outset not only helps to build trust in the evaluation process itself but also allows it to be effective. Moreover, one must once and for all let go of the illusory hope that metrics designed for institutional evaluation will not be copied for the evaluation of individual researchers. Many policy makers truly believe this, but the experience of numerous countries shows that it is not possible. This is because when senior officials reproduce the national solution at the local level, they are acting rationally, according to the logic of their evaluation regime. Thus, the metrics used to evaluate institutions and researchers must be the same.

Fifth: *Let us deindividualize evaluation, that is, let us evaluate researchers as members of research groups, members of departments, or heads of laboratories. In modern science, no one works alone.* This is true even in the humanities, which is frequently presented as the last bastion of single-author articles and monographs. Every researcher develops in an environment with other researchers who build professional relationships, support, and compete with each other. The evaluation of individual researchers on the basis of their scholarly communication reduces their activity to publications. A lab head or research group leader carries out very important social roles in science: supervising the work of others, supporting their development, and coordinating research work. Such a researcher should not be evaluated at his or her institution or even at a grant institution only through the prism of publication numbers. This is the reason for the development of the practice of including lab heads or deans as authors of publications to which they did not contribute, something considered common in medicine, for example. Frequently, this happens not because the supervisor has demanded it but in order to enable them to be held accountable for their work. Thus this is a typical evaluation game: A lab head is expected to run their lab and is held accountable for their publications. Such way of evaluating is a road to nowhere. In addition, viewing science as a collective endeavor will also help restore the role of teaching, which has been degraded in too many universities. It is also important that researchers be judged by the success of their disciples; after all, the success of students says a lot about their master. A researcher who is accountable only for the number of grants and publications they generate ceases to have time for students. And yet the role of science in higher education institutions is not more important than that of teaching.

Sixth: *key scholarly communication infrastructures must be managed by academia itself.* Over the past decades, we have seen key communication channels that were once managed and operated by universities, learned societies, and research institutes, become vehicles for profit generation by private entities. There is no doubt that the academic community has lost control over its journals (Larivière et al., 2015a), and the situation is similar with academic book publishing. It is because major publishers such as Elsevier or Springer-Nature have amassed symbolic capital which is sustained by bibliometric indicators, that researchers are forced by evaluative technologies to publish in journals produced by publishing oligopolies. At the same time, a parallel infrastructure already exists, to an extent, that allows open-access journals to be published by universities, learned societies, and not-for-profit organizations. However, from the perspective of a researcher who wants to keep his or her job, publishing in these parallel channels is not viable. Thus, it is not the lack of technical solutions or infrastructure that are the obstacles to tackling these challenges, but the lack of incentives to publish in channels managed by researchers rather than large corporations. Leadership in this resistance and revolution must be taken by researchers in the higher echelons of the academic hierarchy, as only they have working conditions stable enough to allow them to rebel. Of course, the eagerness and willingness of younger researchers is essential for this change to happen, and young researchers cannot be expected to sacrifice their dreams of working in academia and doing research on the altar of revolution. I chose to describe this process as a revolution, where transformation is rapid and abrupt, because I no longer believe in slow, evolutionary change. The Open Access movement called for gradual and effective change. However, the time that has been dedicated to education and reshaping perceptions about scholarly communication has been cannibalized by the largest commercial players, They have used obligatory open access as another tool for privatizing profits at the expense of publicly funded research.

Seventh: *if metrics are to be part of research evaluation, all data used to calculate them must be completely transparent and accessible to all.* This concerns not only bibliometric indicators but also any metrics and the data used to build them. Many of the negative consequences of the use of bibliometric indicators would be eliminated if the infrastructure for data collection and generation, and not only the infrastructure for scholarly communication, were opened. For example, there are already ventures such as the Initiative for Open Citations, the Initiative for Open Abstracts, or OpenAlex, which are aimed at ultimately providing open access to key bibliometric data for all who wish to view it. Indicators built on this basis will be accurate, which is what the first response wants to achieve. They will also be difficult for large global corporations to take over and shut down, a fact that the first response, which considers giants like Elsevier or Clarivate Analytics as necessary evils, often fails to note. I am quite certain that a future for scholarly communication that does not include Elsevier or Clarivate Analytics (as hegemonic actors) is possible.

References

Aagaard, K. (2015). How incentives trickle down: Local use of a national bibliometric indicator system. *Science and Public Policy, 42*(5), 725–737. https://doi.org/10.1093/scipol/scu087

Aagaard, K. (2018). Performance-based research funding in Denmark: The adoption and translation of the Norwegian model. *Journal of Data and Information Science, 3*(4), 20–30. https://doi.org/10.2478/jdis-2018-0018

Aagaard, K., & Schneider, J. W. (2017). Some considerations about causes and effects in studies of performance-based research funding systems. *Journal of Informetrics, 11*(3), 923–926. https://doi.org/10.1016/j.joi.2017.05.018

Abramo, G., D'Angelo, C. A., & Di Costa, F. (2016). The effect of a country's name in the title of a publication on its visibility and citability. *Scientometrics, 109*(3), 1895–1909. https://doi.org/10.1007/s11192-016-2120-1

Agar, J. (2019). *Science policy under thatcher.* UCL Press. https://doi.org/10.14324/111.9781787353411

Aitkenhead, D. (2013, December 6). Peter Higgs: I wouldn't be productive enough for today's academic system. *The Guardian.* www.theguardian.com/science/2013/dec/06/peter-higgs-boson-academic-system

Allen, A. (2015). *The fantastic laboratory of Dr. Weigl: How two brave scientists battled typhus and sabotaged the Nazis.* W. W. Norton & Company.

Alrawadieh, Z. (2018). Publishing in predatory tourism and hospitality journals: Mapping the academic market and identifying response strategies. *Tourism and Hospitality Research, 20*(1), 72–81. https://doi.org/10.1177/1467358418800121

Andersen, L. B., & Pallesen, T. (2008). "Not just for the money?" How financial incentives affect the number of publications at Danish research institutions. *International Public Management Journal, 11*(1), 28–47. https://doi.org/10.1080/10967490801887889

Andrés, A. (2009). *Measuring academic research: How to undertake a bibliometric study.* Chandos Publishing. https://doi.org/10.1016/B978-1-84334-528-2.50013-8

Angermuller, J., & Van Leeuwen, T. (2019). On the social uses of scientometrics: The quantification of academic evaluation and the rise of numerocracy in higher education. In R. Scholz (Ed.), *Quantifying approaches to discourse for social scientists* (pp. 89–119). Springer International Publishing. https://doi.org/10.1007/978-3-319-97370-8_4

Anisimov, E. V. (1997). *Государственные преобразования и самодержавие Петра Великого в первой четверти XVIII века.* Дмитрий Буланин.

Antonowicz, D., Kohoutek, J., Pinheiro, R., & Hladchenko, M. (2017). The roads of "excellence" in Central and Eastern Europe. *European Educational Research Journal, 16*(5), 547–567. https://doi.org/10.1177/1474904116683186

Antonowicz, D., Kulczycki, E., & Budzanowska, A. (2020). Breaking the deadlock of mistrust? A participative model of the structural reforms in higher education in Poland. *Higher Education Quarterly*, *74*(4), 391–409. https://doi.org/10.1111/hequ.12254

Arnold, E., Simmonds, P., Farla, K., Kolarz, P., Mahieu, B., & Nielsen, K. (2018). *Review of the research excellence framework: Evidence report*. Technopolis Group. https://assets.publishing.service.gov.uk/government/uploads/system/uploads/attachment_data/file/768162/research-excellence-framework-review-evidence-report.pdf

Aronova, E. (2011). The politics and contexts of Soviet science studies (Naukovedenie): Soviet philosophy of science at the crossroads. *Studies in East European Thought*, *63*(3), 175–202. https://doi.org/10.1007/s11212-011-9146-y

Aronova, E. (2021). Scientometrics with and without computers: The cold war transnational journeys of the science citation index. In M. Solovey & C. Dayé (Eds.), *Cold war social science* (pp. 73–98). Palgrave Macmillan.

Avedon, E. M. (1981). The structural elements of games. In A. Furnham & M. Argyle (Eds.), *The psychology of social situations* (pp. 419–426). Pergamon. https://doi.org/10.1016/B978-0-08-023719-0.50009-7

Axtell, J. (2016). *Wisdom's workshop: The rise of the modern university*. Princeton University Press. https://doi.org/10.2307/j.ctv7h0s90

Babones, S., & Babicky, P. (2011, February 3–4). *Russia and East-Central Europe in the modern world-system: A structuralist perspective* [Paper presentation]. Proceedings of the 10th Biennial Conference of the Australasian Association for Communist and Post-Communist Studies, Canberra, Australia. https://core.ac.uk/reader/41235980

Baccini, A., Nicolao, G. D., & Petrovich, E. (2019). Citation gaming induced by bibliometric evaluation: A country-level comparative analysis. *PloS One*, *14*(9), e0221212. https://doi.org/10.1371/journal.pone.0221212

Bacevic, J. (2019). Knowing neoliberalism. *Social Epistemology*, *33*(4), 380–392. https://doi.org/10.1080/02691728.2019.1638990

Bak, H.-J., & Kim, D. H. (2019). The unintended consequences of performance-based incentives on inequality in scientists' research performance. *Science and Public Policy*, *46*(2), 219–231. https://doi.org/10.1093/scipol/scy052

Baker, S. (2019, March 14). UK universities shift to teaching-only contracts ahead of REF. Times Higher Education. www.timeshighereducation.com/news/uk-universities-shift-teaching-only-contracts-ahead-ref

Bal, R. (2017). Playing the indicator game: Reflections on strategies to position an STS group in a multi-disciplinary environment. *Engaging Science, Technology, and Society*, *3*, 41–52. https://doi.org/10.17351/ests2017.111

Baldwin, M. (2018). Scientific autonomy, public accountability, and the rise of "peer review" in the Cold War United States. *Isis*, *109*(3), 538–558. https://doi.org/10.1086/700070

Banaji, M. R. (1998). *William J. McGuire: Remarks offered at the 1998 society of experimental social psychology convention*. https://web.archive.org/web/20210803225509/http://www.people.fas.harvard.edu/~banaji/research/speaking/tributes/mcguire.html

Beall, J. (2012). Predatory publishers are corrupting open access. *Nature*, *489*(7415), 179–179. https://doi.org/10.1038/489179a

Beall, J. (2013). Unethical practices in scholarly, open-access publishing. *Journal of Information Ethics*, *22*(1), 11–20. https://doi.org/10.3172/JIE.22.1.11

Beall, J. (2018). Scientific soundness and the problem of predatory journals. In A. B. Kaufman & J. C. Kaufman (Eds.), *Pseudoscience: The conspiracy against science*. The MIT Press. https://doi.org/10.7551/mitpress/10747.003.0018

Becher, T., & Trowler, P. R. (2001). *Academic tribes and territories: Intellectual enquiry and the culture of disciplines* (2nd ed.). The Society for Research into Higher Education & Open University Press.

Beck, M. T., & Gáspár, V. (1991). Scientometric evaluation of the scientific performance at the Faculty of Natural Sciences, Kossuth Lajos University, Debrecen, Hungary. *Scientometrics*, *20*(1), 37–54. https://doi.org/10.1007/BF02018142

Beer, D. (2016). *Metric power*. Palgrave Macmillan.

Beigel, F. (2021). A multi-scale perspective for assessing publishing circuits in non-hegemonic countries. *Tapuya: Latin American Science, Technology and Society*, *4*(1), 1845923. https://doi.org/10.1080/25729861.2020.1845923

Bence, V., & Oppenheim, C. (2005). The evolution of the UK's research assessment exercise: Publications, performance and perceptions. *Journal of Educational Administration and History*, *37*(2), 137–155. https://doi.org/10.1080/00220620500211189

Berman, E. P. (2014). Not just neoliberalism: Economization in US science and technology policy. *Science, Technology, & Human Values*, *39*(3), 397–431. https://doi.org/10.1177/0162243913509123

Bernal, J. D. (1939). *The social function of science*. George Routledge & Sons. https://doi.org/10.1111/j.1742-1241.2008.01735.x

Bernal, J. D. (1971). *Science in history: Vol. 2. The scientific and industrial revolutions*. The MIT Press.

Bevan, G., & Hood, C. (2007). What's measured is what matters: Targets and gaming in the English public health care system. *Public Administration*, *84*(3), 517–538. https//doi.org/10.1111/j.1467-9299.2006.00600.x

Bhattacharjee, Y. (2011). Saudi universities offer cash in exchange for academic prestige. *Science*, *334*(6061), 1344–1345. https://doi.org/10.1126/science.334.6061.1344

Blackmore, P., & Kandiko, C. B. (2011). Motivation in academic life: A prestige economy. *Research in Post-Compulsory Education*, *16*(4), 399–411. https://doi.org/10.1080/13596748.2011.626971

Blasi, B., Romagnosi, S., & Bonaccorsi, A. (2018). Do SSH researchers have a third mission (and should they have)? In A. Bonaccorsi (Ed.), *The evaluation of research in social sciences and humanities: Lessons from the Italian experience* (pp. 361–392). Springer International Publishing.

Bocanegra-Valle, A. (2019). Building a reputation in global scientific communication: A SWOT analysis of Spanish humanities journals. *Canadian Journal of Sociology*, *44*(1), 39–66. https://doi.org/10.29173/cjs28935

Bohannon, J. (2013). Who's afraid of peer review? *Science*, *342*(6154), 60–65. https://doi.org/10.1126/science.342.6154.60

Bonaccorsi, A. (Ed.). (2018a). *The evaluation of research in social sciences and humanities: Lessons from the Italian experience*. Springer International Publishing. https://doi.org/10.1007/978-3-319-68554-0

Bonaccorsi, A. (2018b). Towards an epistemic approach to evaluation in SSH. In A. Bonaccorsi (Ed.), *The evaluation of research in social sciences and humanities: Lessons from the Italian experience* (pp. 1–29). Springer International Publishing.

Bonev, I. (2009). Should we take journal impact factors seriously? *ParalleMIC*. https://espace2.etsmtl.ca/id/eprint/9919

Bordons, M., Fernández, M. T., & Gómez, I. (2002). Advantages and limitations in the use of impact factor measures for the assessment of research performance in a peripheral country. *Scientometrics*, *53*(2), 195–206. https://doi.org/10.1023/A:1014800407876

Bornmann, L., & Daniel, H.-D. (2008). What do citation counts measure? A review of studies on citing behavior. *Journal of Documentation*, *64*(1), 45–80. https://doi.org/10.1108/00220410810844150

Bornmann, L., & Haunschild, R. (2018). Alternative article-level metrics. *EMBO Reports*, *19*(12), e47260–e47260. https://doi.org/10.15252/embr.201847260

Bornmann, L., & Leydesdorff, L. (2014). Scientometrics in a changing research landscape: Bibliometrics has become an integral part of research quality evaluation and has been changing the practice of research. *EMBO Reports, 15*(12), 1228–1232. https://doi.org/10.15252/embr.201439608

Bornmann, L., & Williams, R. (2017). Can the journal impact factor be used as a criterion for the selection of junior researchers? A large-scale empirical study based on ResearcherID data. *Journal of Informetrics, 11*(3), 788–799. https://doi.org/10.1016/j.joi.2017.06.001

Bourdieu, P., & Wacquant, L. J. D. (1992). *An invitation to reflexive sociology.* Polity Press.

Bowman, J. D. (2014). Predatory publishing, questionable peer review, and fraudulent conferences. *American Journal of Pharmaceutical Education, 78*(10), 176. https://doi.org/10.5688/ajpe7810176

Bradford, S. C. (1948). *Documentation.* Crosby Lockwood and Son.

Braudel, F. (1958). Histoire et Sciences sociales: La longue durée. *Annales. Histoire, Sciences Sociales, 13*(4), 725–753. https://doi.org/10.3406/ahess.1958.2781

Brož, L., Stöckelová, T., & Vostal, F. (2017, January 26). *Predators and bloodsuckers in academic publishing.* Derivace. https://derivace.wordpress.com/2017/01/26/predators-and-bloodsuckers-in-academic-publishing/

Buranyi, S. (2017, June 27). Is the staggeringly profitable business of scientific publishing bad for science? *The Guardian.* www.theguardian.com/science/2017/jun/27/profitable-business-scientific-publishing-bad-for-science

Burton, M. (2016). *The politics of austerity: A recent history.* Springer.

Butler, L. (2003a). Explaining Australia's increased share of ISI publications: The effects of a funding formula based on publication counts. *Research Policy, 32*(1), 143–155. https://doi.org/10.1016/s0048-7333(02)00007-0

Butler, L. (2003b). Modifying publication practices in response to funding formulas. *Research Evaluation, 12*(1), 39–46. https://doi.org/10.3152/147154403781776780

Butler, L. (2017). Response to van den Besselaar et al.: What happens when the Australian context is misunderstood. *Journal of Informetrics, 11*(3), 919–922. https://doi.org/10.1016/j.joi.2017.05.017

Cai, Y. (2010). Global isomorphism and governance reform in Chinese higher education. *Tertiary Education and Management, 16*(3), 229–241. https://doi.org/10.1080/13583883.2010.497391

Campbell, D. T. (1979). Assessing the impact of planned social change. *Evaluation and Program Planning, 2*(1), 67–90.

Chapelle, F. H. (2014). The history and practice of peer review. *Groundwater, 52*(1), 1. https://doi.org/10.1111/gwat.12139

Chapman, C. A., Bicca-Marques, J. C., Calvignac-Spencer, S., Fan, P., Fashing, P. J., Gogarten, J., Guo, S., Hemingway, C. A., Leendertz, F., Li, B., Matsuda, I., Hou, R., Serio-Silva, J. C., & Chr. Stenseth, N. (2019). Games academics play and their consequences: How authorship, h-index and journal impact factors are shaping the future of academia. *Proceedings of the Royal Society B: Biological Sciences, 286*(1916), 20192047. https://doi.org/10.1098/rspb.2019.2047

Chavarro, D., Tang, P., & Ràfols, I. (2017). Why researchers publish in non-mainstream journals: Training, knowledge bridging, and gap filling. *Research Policy, 46*(9), 1666–1680. https://doi.org/10.1016/j.respol.2017.08.002

Chen, X. (2019). High monetary rewards and high academic article outputs: Are China's research publications policy driven? *The Serials Librarian, 77*(1–2), 49–59. https://doi.org/10.1080/0361526X.2019.1645793

Chung, A. (2018). *Media probes raise questions over quality of conferences.* University World News. www.universityworldnews.com/post.php?story=20180922053255197

Cleere, L., & Ma, L. (2018). A local adaptation in an Output-Based Research Support Scheme (OBRSS) at University College Dublin. *Journal of Data and Information Science*, *3*(4), 74–84. https://doi.org/10.2478/jdis-2018-0022

Cocka, P. M. (1980). *Science policy: USA/USSR. Volume II, science policy in the Soviet Union*. National Science Foundation. https://files.eric.ed.gov/fulltext/ED199102.pdf

Connelly, J. (2000). *Captive university: The Sovietization of East German, Czech and Polish higher education, 1945–1956*. University of North Carolina Press.

Cooper, S., & Poletti, A. (2011). The new ERA of journal ranking: The consequences of Australia's fraught encounter with "quality." *Australian Universities Review*, *53*(1), 57–65.

Crocker, J., & Cooper, M. L. (2011). Addressing scientific fraud. *Science*, *334*(6060), 1182–1182. https://doi.org/10.1126/science.1216775

Cronin, B. (2015). The need for a theory of citing. In B. Cronin & C. R. Sugimoto (Eds.), *Scholarly metrics under the microscope: From citation analysis to academic auditing* (pp. 33–44). Information Today. https://doi.org/10.1108/eb026703

Cronin, B., & Sugimoto, C. R. (Eds.). (2014). *Beyond bibliometrics: Harnessing multidimensional indicators of scholarly impact*. The MIT Press.

Dadkhah, M., Borchardt, G., Lagzian, M., & Bianciardi, G. (2017). Academic journals plagued by bogus impact factors. *Publishing Research Quarterly*, *33*(2), 183–187. https://doi.org/10.1007/s12109-017-9509-4

Dahler-Larsen, P. (2012). *The evaluation society*. Stanford Business Books.

Dahler-Larsen, P. (2014). Constitutive effects of performance indicators: Getting beyond unintended consequences. *Public Management Review*, *16*(7), 969–986. https://doi.org/10.1080/14719037.2013.770058

Dahler-Larsen, P. (2015). The evaluation society: Critique, contestability and skepticism. *SpazioFilosofico*, *1*(13), 21–36. www.spaziofilosofico.it/numero-13/5241/the-evaluation-society-critique-contestability-and-skepticism/

Dahler-Larsen, P. (2022). Your brother's gatekeeper: How effects of evaluation machineries in research are sometimes enhanced. In E. Forsberg, L. Geschwind, S. Levander, & W. Wermke (Eds.), *Peer review in an era of evaluation* (pp. 127–146). Palgrave Macmillan.

Davis, K. E., Fisher, A., Kingsbury, B., & Merry, S. E. (Eds.). (2012). *Governance by indicators: Global power through quantification and rankings*. Oxford University Press.

Davis, K. E., Kingsbury, B., & Merry, S. (2012). Introduction: Global governance by indicators. In K. E. Davis, A. Fisher, B. Kingsbury, & S. E. Merry (Eds.), *Governance by indicators: Global power through quantification and rankings* (pp. 3–28). Oxford University Press.

Davis, P. (2016, February 10). *Citable items: The contested impact factor denominator*. The Scholarly Kitchen. https://scholarlykitchen.sspnet.org/2016/02/10/citable-items-the-contested-impact-factor-denominator/

De Bellis, N. (2009). *Bibliometrics and citation analysis: From the science citation index to cybermetrics*. The Scarecrow Press.

De Bellis, N. (2014). History and evolution of (biblio)metrics. In B. Cronin & C. R. Sugimoto (Eds.), *Beyond bibliometrics: Harnessing multidimensional indicators of scholarly impact* (pp. 23–44). The MIT Press.

De Jong, S. P. L., & Muhonen, R. (2018). Who benefits from ex ante societal impact evaluation in the European funding arena? A cross-country comparison of societal impact capacity in the social sciences and humanities. *Research Evaluation*, *29*(1), 1–12. https://doi.org/10.1093/reseval/rvy036

De Rijcke, S., & Stöckelová, T. (2020). Predatory publishing and the imperative of international productivity: Feeding off and feeding up the dominant. In M. Biagioli & A. Lippman (Eds.), *Gaming the metrics: Misconduct and manipulation in academic research* (pp. 101–110). MIT Press.

De Rijcke, S., Wouters, P. F., Rushforth, A. D., Franssen, T. P., & Hammarfelt, B. (2016). Evaluation practices and effects of indicator use: A literature review. *Research Evaluation*, *25*(2), 161–169. https://doi.org/10.1093/reseval/rvv038

Demir, S. B. (2018). Predatory journals: Who publishes in them and why? *Journal of Informetrics*, *12*(4), 1296–1311. https://doi.org/10.1016/j.joi.2018.10.008

Derrick, G. (2018). *The evaluators' eye: Impact assessment and academic peer review.* Palgrave Macmillan.

Desrosières, A. (2002). *The politics of large numbers: A history of statistical reasoning* (C. Naish, Trans.). Harvard University Press.

Desrosières, A. (2013). *Pour une sociologie historique de la quantification: L'Argument statistique I.* Presses des Mines. https://doi.org/10.4000/books.pressesmines.901

Dey, E. L., Milem, J. F., & Berger, J. B. (1997). Changing patterns of publication productivity: Accumulative advantage or institutional isomorphism? *Sociology of Education*, *70*(4), 308–323. https://doi.org/10.2307/2673269

Dill, D. D. (2014). Evaluating the "evaluative state": Implications for research in higher education. *European Journal of Education*, *33*(3), 361–377.

DiMaggio, P. J., & Powell, W. W. (1983). The iron cage revisited: Institutional isomorphism and collective rationality in organizational fields. *American Sociological Review*, *48*(2), 147–160. https://doi.org/10.2307/2095101

Directorate-General for Research and Innovation. (2014). *Innovation Union Competitiveness report: Commission Staff Working Document.* https://op.europa.eu/en/publication-detail/-/publication/799d9836-1333-4804-835a-6968b35ae619

Dmitriev, I. S. (2016). Академия благих надежд (эффективность научной деятельности Петербургской академии наук в XVIII столетии). *Социология Науки и Технологий*, *7*(4). 9–31.

Dobija, D., Górska, A. M., & Pikos, A. (2019). The impact of accreditation agencies and other powerful stakeholders on the performance measurement in Polish universities. *Baltic Journal of Management*, *14*(1), 84–102. https://doi.org/10.1108/BJM-01-2018-0018

Dobrov, G. (1969a). Kryterium wyboru jako kompleksowy problem naukoznawstwa. *Zagadnienia Naukoznawstwa*, *5*(4), 92–99.

Dobrov, G. (1969b). *Wstęp do naukoznawstwa* (J. Bolecki, Trans.). Państwowe Wydawnictwo Naukowe.

Donovan, C. (2008). The Australian research quality framework: A live experiment in capturing the social, economic, environmental, and cultural returns of publicly funded research. *New Directions for Evaluation*, *2008*(118), 47–60. https://doi.org/10.1002/ev.260

Drabek, A., Rozkosz, E., & Kulczycki, E. (2017). *Analiza opóźnień wydawniczych polskich czasopism naukowych* (Report No. 3). https://doi.org/10.6084/M9.FIGSHARE.4578394

Dunleavy, P., & Hood, C. (1994). From old public administration to new public management. *Public Money and Management*, *14*(3), 9–16. https://doi.org/10.1080/09540969409387823

Eastwood, D. (2007). Goodbye to the RAE … And hello to the REF. *Times Higher Education*. www.timeshighereducation.com/news/goodbye-to-the-rae-and-hello-to-the-ref-david-eastwood/311322.article

Eisenstadt, S. N. (2000). Multiple modernities. *Daedalus*, *129*(1), 1–29.

Ellman, M. (2014). *Socialist planning* (3rd ed.). Cambridge University Press.

Elsaie, M., & Kammer, J. (2009). Impactitis: The impact factor myth syndrome. *Indian Journal of Dermatology*, *54*(1), 83–86. https://doi.org/10.4103/0019-5154.48998

Else, H. (2019, April 11). Impact factors are still widely used in academic evaluations. *Nature*. https://doi.org/10.1038/d41586-019-01151-4

Else, H. (2022, March 14). Journals under pressure to boycott Russian authors. *Nature*. https://doi.org/10.1038/d41586-022-00718-y

Emysheva, E. M. (2008). Генеральный регламент 1720 года как опыт создания организационного документа. *История и Архивы, 8*, 248–261.

Enders, J. (2001). A chair system in transition: Appointments, promotions, and gate-keeping in German higher education. *Higher Education, 41*(1–2), 3–25. https://doi .org/10.1023/A:1026790026117

Engels, T. C. E., & Guns, R. (2018). The Flemish performance-based research funding system: A unique variant of the Norwegian model. *Journal of Data and Information Science, 3*(4), 45–60. https://doi.org/10.2478/jdis-2018-0020

Engels, T. C. E., Isteniĉ Starĉiĉ, A., Kulczycki, E., Pölönen, J., & Sivertsen, G. (2018). Are book publications disappearing from scholarly communication in the social sciences and humanities? *Aslib Journal of Information Management, 70*(6), 592–607. https://doi .org/10.1108/AJIM-05-2018-0127

Erfanmanesh, M., & Pourhossein, R. (2017). Publishing in predatory open access journals: A case of Iran. *Publishing Research Quarterly, 33*(4), 433–444. https://doi.org/10.1007/ s12109-017-9547-y

Espeland, W. N., & Sauder, M. (2007). Rankings and reactivity: How public measures recreate social worlds. *American Journal of Sociology, 113*(1), 1–40. https://doi .org/10.1086/517897

Espeland, W. N., & Stevens, M. L. (2008). A sociology of quantification. *European Journal of Sociology/Archives Européennes de Sociologie, 49*(3), 401–436. https://doi .org/10.1017/S0003975609000150

Etzkowitz, H., & Leydesdorff, L. (2000). The dynamics of innovation: From national systems and "mode 2" to a triple helix of university–industry–government relations. *Research Policy, 29*(2), 109–123. https://doi.org/10.1016/S0048-7333(99)00055-4

Ezinwa Nwagwu, W., & Ojemeni, O. (2015). Penetration of Nigerian predatory biomedical open access journals 2007–2012: A bibliometric study. *Learned Publishing, 28*(1), 23–34. https://doi.org/10.1087/20150105

Fernández, L. M., & Vadillo, M. A. (2019, June 19). *Retracted papers die hard: Diederik Stapel and the enduring influence of flawed science*. https://doi.org/10.31234/osf.io/ cszpy

Fochler, M., & De Rijcke, S. (2017). Implicated in the indicator game? An experimental debate. *Engaging Science, Technology, and Society, 3*(3), 21–40. https://doi .org/10.17351/ests2017.108

Foucault, M. (1982). The subject and power. *Critical Inquiry, 8*(4), 777–795.

Foucault, M. (1995). *Discipline & punish: The birth of the prison*. Vintage Books.

Frankel, M. S., & Cave, J. (Eds.). (1997). *Evaluating science and scientists: An east-west dialogue on research evaluation in post-communist Europe*. Central European University Press.

French, N. J., Massy, W. F., & Young, K. (2001). Research assessment in Hong Kong. *Higher Education, 42*, 34–46.

Frood, A. (2015, February 9). Death in academia and the mis-measurement of science. EuroScientist. www.euroscientist.com/death-academia-mis-measurement-science/

Fyfe, A., Squazzoni, F., Torny, D., & Dondio, P. (2019). Managing the growth of peer review at the Royal Society Journals, 1865–1965. *Science, Technology, & Human Values, 45*(3), 405–429. https://doi.org/10.1177/0162243919862868

Galiullina, R. H., & Ilina, K. A. (2012). «Ученые записки» ученого сословия (первая половина XIX века). *Препринты. Высшая Школа Экономики. WP6/2012/02 Серия WP6 Гуманитарные исследования*

Garfield, E. (1955). Citation indexes for science: A new dimension in documentation through association of ideas. *Science*, *122*(3159), 108–111. https://doi.org/10.1126/science .122.3159.108

Garfield, E. (1964). Can citation indexing be automated? In M. E. Stevens, V. E. Giuliano, & L. B. Heilprin (Eds.), *Statistical association methods for mechanized documentation: Symposium proceedings* (pp. 189–192). National Bureau of Standards.

Garfield, E. (1996). Significant scientific literature appears in a small core of journals. *The Scientist*, *10*(17), 13–15.

Garfield, E. (2005, September 16). *The agony and the ecstasy: The history and meaning of the journal impact factor* [Paper presentation]. International Congress on Peer Review and Biomedical Publication, Chicago, IL, United States. http://garfield.library.upenn .edu/papers/jifchicago2005.pdf

Garfield, E. (2006). The history and meaning of the journal impact factor. *JAMA*, *295*(1), 90–93. https://doi.org/10.1001/jama.295.1.90

Geuna, A., & Martin, B. R. (2003). University research evaluation and funding: An international comparison. *Minerva*, *41*(4), 277–304. https://doi.org/10.1023/B:MINE .0000005155.70870.bd

Giménez Toledo, E. (2016). *Malestar: Los investigadores ante su evaluación*. Iberoamericana.

Gingras, Y. (1995, September 13–15). *Performance indicators: Keeping the black box open* [Paper Presentation]. Dans Proceedings of the Second International Symposium on Research Funding, Ottawa, Canada. https://ost.openum.ca/files/sites/132/2017/06/ proceedings_research_funding.pdf

Gingras, Y. (2014). *Bibliometrics and research evaluation: Uses and abuses*. The MIT Press.

Glänzel, W., Moed, H. F., Schmoch, U., & Thelwall, M. (Eds.). (2019). *Handbook of science and technology indicators*. Springer. https://doi.org/10.1007/978-3-030-02511-3

Gläser, J., Lange, S., Laudel, G., & Schimank, U. (2010). Informed authority? The limited use of research evaluation systems for managerial control in universities. In R. Whitley, J. Gläser, & L. Engwall (Eds.), *Reconfiguring knowledge production: Changing authority relationships in the sciences and their consequences for intellectual innovation* (pp. 149–183). Oxford University Press.

Gläser, J., & Laudel, G. (2007). The social construction of bibliometric evaluations. In R. Whitley & J. Gläser (Eds.), *The changing governance of the sciences: The advent of research evaluation systems* (pp. 101–123). Springer.

Godin, B. (2005). *Measurement and statistics on science and technology: 1920 to the present*. Routledge Taylor & Francis Group. https://doi.org/10.1016/j.spinee.2013.10.043

Godin, B. (2006). From eugenics to scientometrics: Galton, Cattell and men of science. *Social Studies of Science*, *37*(5), 691–728. https://doi.org/10.1177/0306312706075338

Godin, B. (2009). The value of science: Changing conceptions of scientific productivity, 1869 to circa 1970. *Social Science Information*, *48*(4), 547–586. https://doi .org/10.1177/0539018409344475

Goffman, E. (1972). *Encounters: Two studies in the sociology of interaction*. Penguin University Books.

Good, B., Vermeulen, N., Tiefenthaler, B., & Arnold, E. (2015). Counting quality? The Czech performance-based research funding system. *Research Evaluation*, *24*(2), 91–105. https://doi.org/10.1093/reseval/rvu035

Graham, L. R. (1964). Bukharin and the planning of Science. *Russian Review*, *23*(2), 135–148. https://doi.org/10.2307/126518

Graham, L. R. (1967). *The Soviet academy of sciences and the communist party, 1927–1932*. Princeton University Press.

Graham, L. R. (Ed.). (1990). *Science and the Soviet social order*. Harvard University Press.

Graham, L. R. (1993). *Science in Russia and the Soviet Union: A short history*. Cambridge University Press.

Graham, L. R., & Dezhina, I. (2008). *Science in the new Russia: Crisis, aid, reform*. Indiana University Press.

Granovsky, Y. V. (2001). Is it possible to measure science? V. V. Nalimov's research in scientometrics. *Scientometrics, 52*(2), 127–150. https://doi.org/10.1023/A:1017991017982

Grzechnik, M. (2019). The missing second world: On Poland and postcolonial studies. *Interventions, 21*(7), 998–1014. https://doi.org/10.1080/1369801X.2019.1585911

Guskov, A., Kosyakov, D., & Selivanova, I. (2016). Scientometric research in Russia: Impact of science policy changes. *Scientometrics, 107*(1), 287–303. https://doi.org/10.1007/s11192-016-1876-7

Gutierrez, F. R. S., Beall, J., & Forero, D. A. (2015). Spurious alternative impact factors: The scale of the problem from an academic perspective. *BioEssays, 37*(5), 474–476. https://doi.org/10.1002/bies.201500011

Haddow, G. (2022). Research assessment in Australia: Journal ranking, research classification and ratings. In T. C. E. Engels & E. Kulczycki (Eds.), *Handbook on research assessment in the social sciences* (pp. 434–450). Edward Elgar Publishing.

Haitun, S. D. (1980). Scientometric investigations in the USSR: Review. *Scientometrics, 2*(1), 65–84. https://doi.org/10.1007/BF02016600

Hallonsten, O. (2021). Stop evaluating science: A historical-sociological argument. *Social Science Information, 60*(1), 7–26. https://doi.org/10.1177/0539018421992204

Hammarfelt, B. (2018). Taking comfort in points: The appeal of the Norwegian model in Sweden. *Journal of Data and Information Science, 3*(4), 85–95. https://doi.org/10.2478/jdis-2018-0023

Hammarfelt, B., & Åström, F. (2015, September 2–4). The multi-layered and multilevel use of bibliometric measures in Swedish universities: Isomorphism, translation and strategic choice [Paper Presentation]. The 20th International Conference on Science and Technology Indicators, Lugano, Switzerland. http://hb.diva-portal.org/smash/get/diva2:853400/FULLTEXT01.pdf

Hammarfelt, B., & Haddow, G. (2018). Conflicting measures and values: How humanities scholars in Australia and Sweden use and react to bibliometric indicators. *Journal of the Association for Information Science and Technology, 69*(7), 924–935. https://doi.org/10.1002/asi.24043

Hanley, H. J. M. (1975). Letters to the editor. *Science, 188*(4193), 1064. https://doi.org/10.1126/science.188.4193.1064-a

Hare, P. (1989). The economics of shortage in the centrally planned economies. In C. Davis & W. Charemza (Eds.), *Models of disequilibrium and shortage in centrally planned economies* (pp. 49–81). Springer Netherlands. https://doi.org/10.1007/978-94-009-0823-9_3

Hare, P. (1991). *Central planning*. Routledge Taylor & Francis Group.

Harvey, D. (2005). *A brief history of neoliberalism*. Oxford University Press.

Hazelkorn, E. (2015). *Rankings and the reshaping of higher education: The battle for world-class excellence* (2nd ed.). Palgrave Macmillan.

Hedding, D. W. (2019). Payouts push professors towards predatory journals. *Nature, 565*(7739), 267–267. https://doi.org/10.1038/d41586-019-00120-1

Hicks, D. (2012). Performance-based university research funding systems. *Research Policy, 41*(2), 251–261. https://doi.org/10.1016/j.respol.2011.09.007

Hicks, D. (2017). What year? Difficulties in identifying the effect of policy on university output. *Journal of Informetrics, 11*(3), 933–936. https://doi.org/10.1016/j.joi.2017.05.020

Hicks, D., Wouters, P., Waltman, L., De Rijcke, S., & Rafols, I. (2015). Bibliometrics: The Leiden manifesto for research metrics. *Nature, 520*(7548), 429–431. https://doi.org/10.1038/520429a

Higher Education Funding Council for England. (2015). The metric tide: Correlation analysis of REF2014 scores and metrics (Supplementary Report II to the Independent Review of the Role of Metrics in Research Assessment and Management). Higher Education Funding Council for England. https://doi.org/10.13140/RG.2.1.3362.4162

Hilton, A. (2013, April 12). Honoris causa? Margaret Thatcher and the eternal shame of Oxford University. Mail Online. www.dailymail.co.uk/columnists/index.html

Hirsch, J. E. (2005). An index to quantify an individual's scientific research output. *Proceedings of the National Academy of Sciences, 102*(46), 16569–16572. https://doi.org/10.1073/pnas.0507655102

Hood, C. (1991). A public management for all seasons? *Public Administration, 69*(1), 3–19. https://doi.org/10.1111/j.1467-9299.1991.tb00779.x

Horta, H., & Shen, W. (2020). Current and future challenges of the Chinese research system. *Journal of Higher Education Policy and Management, 42*(2), 157–177. https://doi.org/10.1080/1360080X.2019.1632162

Howarth, P. (2022, March 4). Ukraine crisis: A message from THE's chief executive. Times Higher Education. www.timeshighereducation.com/ukraine-crisis-message-thes-chief-executive

Hrabak, M., Vujaklija, A., Vodopivec, I., Hren, D., Marusic, M., & Marusic, A. (2004). Academic misconduct among medical students in a post-communist country. *Medical Education, 38*(3), 276–285. https://doi.org/10.1111/j.1365-2923.2004.01766.x

Huang, Y., Li, R., Zhang, L., & Sivertsen, G. (2020). A comprehensive analysis of the journal evaluation system in China. *Quantitative Science Studies, 2*(1), 300–326. https://doi.org/10.1162/qss_a_00103

Hur, J. D., & Nordgren, L. F. (2016). Paying for performance: Performance incentives increase desire for the reward object. *Journal of Personality and Social Psychology, 111*(3), 301–316. https://doi.org/10.1037/pspa0000059

Hur, J., Lee-Yoon, A., & Whillans, A. V. (2018). *Who is more useful? The impact of performance incentives on work and personal relationships.* Harvard Business School.

Igami, M., & Nagaoka, S. (2014). Exploring the effects of the motivation of a research project on the research team composition, management, and outputs. In E. Noyons (Ed.), *Proceedings of the science and technology indicators conference 2014 Leiden "Context Counts: Pathways to master big and little data"* (pp. 290–294). Universiteit Leiden.

Ings, S. (2016). *Stalin and the scientists: A history of triumph and tragedy 1905–1953.* Grove Press.

Irwin, A. (2017). If the indicator game is the answer, then what is the question? *Engaging Science, Technology, and Society, 3*, 64–72. https://doi.org/10.17351/ests2017.110

Jablecka, J. (1997). Peer review in Poland: Practical solutions and possible improvements. In M. S. Frankel, & J. Cave (Eds.), *Evaluating science and scientists* (pp. 96–111). Central European University Press.

Jacques, T. S., & Sebire, N. J. (2010). The impact of article titles on citation hits: An analysis of general and specialist medical journals. *JRSM Short Reports, 1*(1), 1–5. https://doi.org/10.1258/shorts.2009.100020

Jessop, B. (2002). Globalization and the national state. In S. Aronowitz & P. Bratsis (Eds.), *Paradigm lost: State theory reconsidered* (pp. 185–220). University of Minnesota Press. https://doi.org/10.5749/j.ctttsh78

Jessop, B. (2005). Cultural political economy, the knowledge-based economy, and the state. In A. Barry & D. Slater (Eds.), *Technological economy* (pp. 144–165). Routledge. https://doi.org/10.4324/9780203022450

Jessop, B. (2017). Varieties of academic capitalism and entrepreneurial universities: On past research and three thought experiments. *Higher Education, 73*(6), 853–870. https://doi.org/10.1007/s10734-017-0120-6

Jin, B., & Rousseau, R. (2005). Evaluation of research performance and scientometric indicators in China. In H. F. Moed, W. Glänzel, & U. Schmoch (Eds.), *Handbook of quantitative science and technology research: The use of publication and patent statistics in studies of S&T systems* (pp. 497–514). Kluwer Academic Publishers. https://doi.org/10.1007/1-4020-2755-9_23

Johnson, J. (2017). *Sputnik and the space race.* Cavendish Square Publishing.

Josephson, P. R. (1992). Soviet scientists and the state: Politics, ideology, and fundamental research from Stalin to Gorbachev. *Social Research, 59*(3), 589–614.

Jump, P. (2013, September 26). Twenty per cent contracts rise in run-up to REF. *Times Higher Education.* www.timeshighereducation.com/news/twenty-per-cent-contracts-rise-in-run-up-to-ref/2007670.article

Jump, P. (2015, January 1). REF 2014 rerun: Who are the "game players"? *Times Higher Education.* www.timeshighereducation.com/features/ref-2014-rerun-who-are-the-game-players/2017670.article

Kalfa, S., Wilkinson, A., & Gollan, P. J. (2018). The academic game: Compliance and resistance in universities. *Work, Employment and Society, 32*(2), 274–291. https://doi.org/10.1177/0950017017695043

Kapitza, P. L. (1966). Scientific policy in the U.S.S.R.: The scientist and the plans. *Minerva, 4*(4), 555–560.

Kassian, A., & Melikhova, L. (2019). Russian science citation index on the WoS platform: A critical assessment. *Journal of Documentation, 75*(5), 1162–1168. https://doi.org/10.1108/JD-02-2019-0033

Kehm, B. M. (2014). Global university rankings: Impacts and unintended side effects. *European Journal of Education, 49*(1), 102–112. https://doi.org/10.1111/ejed.12064

Killian, J. R. (1977). *Sputnik, scientists and Eisenhower: A memoir of the first special assistant to the president for science and technology.* MIT Press.

Kim, D. H., & Bak, H.-J. (2016). How do scientists respond to performance-based incentives? Evidence from South Korea. *International Public Management Journal, 19*(1), 31–52. https://doi.org/10.1080/10967494.2015.1032460

King, R., Marginson, S., & Naidoo, R. (Eds.). (2011). *Handbook on globalization and higher education.* Edward Elgar.

Kliucharev, G. A., & Neverov, A. V. (2018). Project "5–100": Some interim results. *RUDN Journal of Sociology, 18*(1), 100–116. https://doi.org/10.22363/2313-2272-2018-18-1-100-116

Koçak, Z. (2020). Precise and immediate action against predatory conferences. *Balkan Medical Journal, 37*(1), 1–2. https://doi.org/10.4274/balkanmedj.galenos.2020.2020.1.001

Kokowski, M. (2015). The science of science (Naukoznawstwo) in Poland: The changing theoretical perspectives and political contexts: A historical sketch from the 1910s to 1993. *Organon, 47*, 147–237.

Kokowski, M. (2016). The science of science (naukoznawstwo) in Poland: Defending and removing the past in the cold war. In E. Aronova & S. Turchetti (Eds.), *Science studies during the cold war and beyond* (pp. 149–176). Palgrave Macmillan US. https://doi.org/10.1057/978-1-137-55943-2_7

Korytkowski, P., & Kulczycki, E. (2019). Publication counting methods for a national research evaluation exercise. *Journal of Informetrics, 13*(3), 804–816. https://doi.org/10.1016/j.joi.2019.07.001

Korytkowski, P., & Kulczycki, E. (2021). The gap between Plan S requirements and grantees' publication practices. *Journal of Informetrics, 15*(2), 101156. https://doi.org/10.1016/j.joi.2021.101156

Kosyakov, D., & Guskov, A. (2019). Research assessment and evaluation in Russian fundamental science. *Procedia Computer Science*, *146*, 11–19. https://doi.org/10.1016/j .procs.2019.01.072

Koza, Z., Lew, R., Kulczycki, E., & Stec, P. (in press). Who controls the national academic promotion system: An analysis of power distribution in Poland.

Krawczyk, F., & Kulczycki, E. (2021a). How is open access accused of being predatory? The impact of Beall's lists of predatory journals on academic publishing. *The Journal of Academic Librarianship*, *47*(2), 1–11. https://doi.org/10.1016/j.acalib.2020.102271

Krawczyk, F., & Kulczycki, E. (2021b). On the geopolitics of academic publishing: The mislocated centers of scholarly communication. *Tapuya: Latin American Science, Technology and Society*, *4*(1), 1984641. https://doi.org/10.1080/25729861.2021.19 84641

Krawczyk, S., Szadkowski, K., & Kulczycki, E. (2021). Identifying top researchers in highly metricized academia: Two discursive strategies of senior officials in Poland. *Discourse: Studies in the Cultural Politics of Education*, 1–12. https://doi.org/10.1080/ 01596306.2021.1993792

Krzeski, J., Szadkowski, K., & Kulczycki, E. (2022). Creating evaluative homogeneity: Experience of constructing a national journal ranking. *Research Evaluation*, rvac011. https://doi.org/10.1093/reseval/rvac011

Kula, W. (1986). *Measures and men*. Princeton University Press.

Kulczycki, E. (2018). The diversity of monographs: Changing landscape of book evaluation in Poland. *Aslib Journal of Information Management*, *70*(6), 608–622. https://doi .org/10.1108/AJIM-03-2018-0062

Kulczycki, E. (2019). Field patterns of scientometric indicators use for presenting research portfolio for assessment. *Research Evaluation*, *28*(2), 169–181. https://doi.org/10.1093/ reseval/rvy043

Kulczycki, E., Engels, T. C. E., Pölönen, J., Bruun, K., Dušková, M., Guns, R., Nowotniak, R., Petr, M., Sivertsen, G., Istenič Starčič, A., & Zuccala, A. (2018). Publication patterns in the social sciences and humanities: Evidence from eight European countries. *Scientometrics*, *116*(1), 463–486. https://doi.org/10.1007/s11192-018-2711-0

Kulczycki, E., Guns, R., Pölönen, J., Engels, T. C. E., Rozkosz, E. A., Zuccala, A. A., Bruun, K., Eskola, O., Starčič, A. I., Petr, M., & Sivertsen, G. (2020). Multilingual publishing in the social sciences and humanities: A seven-country European study. *Journal of the Association for Information Science and Technology*, *71*(11), 1371–1385. https:// doi.org/10.1002/asi.24336

Kulczycki, E., Hołowiecki, M., Taşkın, Z., & Doğan, G. (2022). Questionable conferences and presenters from top-ranked universities. *Journal of Information Science*. https://doi .org/10.1177/01655515221087674

Kulczycki, E., Hołowiecki, M., Taşkın, Z., & Krawczyk, F. (2021a). Citation patterns between impact-factor and questionable journals. *Scientometrics*, *126*(4), 8541–8560. https://doi.org/10.1007/s11192-021-04121-8

Kulczycki, E., Huang, Y., Zucalla, A. A., Engels, T. C. E., Ferrara, A., Guns, R., Pölönen, J., Sivertsen, G., Taşkin, Z., & Zhang, L. (2022). Uses of the Journal Impact Factor in national journal rankings in China and Europe. *Journal of the Association for Information Science and Technology*, *73*(12), 1741–1754. https://doi.org/10.1002/asi.24706

Kulczycki, E., & Korytkowski, P. (2018). Redesigning the model of book evaluation in the polish performance-based research funding system. *Journal of Data and Information Science*, *3*(4), 60–72. https://doi.org/10.2478/jdis-2018-0021

Kulczycki, E., Korzeń, M., & Korytkowski, P. (2017). Toward an excellence-based research funding system: Evidence from Poland. *Journal of Informetrics*, *11*(1), 282–298. https:// doi.org/10.1016/j.joi.2017.01.001

Kulczycki, E., & Rotnicka, S. (2022, September 7–9). *Consequences of participating in questionable academia: A global survey of authors of journal articles and conference presentations* [Paper Presentation]. 26th International Conference on Science and Technology Indicators, Granada, Spain. https://doi.org/10.5281/zenodo.6960060

Kulczycki, E., & Rozkosz, E. A. (2017). Does an expert-based evaluation allow us to go beyond the Impact Factor? Experiences from building a ranking of national journals in Poland. *Scientometrics, 111*(1), 417–442. https://doi.org/10.1007/s11192-017-2261-x

Kulczycki, E., Rozkosz, E. A., & Drabek, A. (2015). Publikacje polskich badaczy w czasopismach z list ERIH w kontekście ewaluacji jednostek naukowych. *Kultura i Edukacja, 24*(1), 149–172. https://doi.org/10.15804/kie.2015.01.08

Kulczycki, E., Rozkosz, E. A., & Drabek, A. (2019a). Internationalization of Polish journals in the social sciences and humanities: Transformative role of the research evaluation system. *Canadian Journal of Sociology, 44*(1), 9–30. https://doi.org/10.29173/cjs28794

Kulczycki, E., Rozkosz, E. A., Engels, T. C. E., Guns, R., & Hołowiecki, M. (2019b). How to identify peer-reviewed publications: Open-identity labels in scholarly book publishing. *PLoS ONE, 14*(3), e0214423. https://doi.org/10.1371/journal.pone.0214423

Kulczycki, E., Rozkosz, E. A., Szadkowski, K., Ciereszko, K., Hołowiecki, M., & Krawczyk, F. (2021b). Local use of metrics for the research assessment of academics: The case of Poland. *Journal of Higher Education Policy and Management, 43*(4), 1–19. https://doi.org/10.1080/1360080X.2020.1846243

Kurt, S. (2018). Why do authors publish in predatory journals? *Learned Publishing, 31*(2), 141–147. https://doi.org/10.1002/leap.1150

Kuzhabekova, A. (2019). Invisibilizing Eurasia: How north–south dichotomization marginalizes post-Soviet scholars in international research collaborations. *Journal of Studies in International Education, 24*(1), 113–130. https://doi.org/10.1177/1028315319888887

Lakhotia, S. C. (2017). Mis-conceived and mis-implemented academic assessment rules underlie the scourge of predatory journals and conferences. *Proceedings of the Indian National Science Academy, 83*(3), 513–515. https://doi.org/10.16943/ptinsa/2017/49141

Lang, R., Mintz, M., Krentz, H. B., & Gill, M. J. (2018). An approach to conference selection and evaluation: Advice to avoid "predatory" conferences. *Scientometrics, 118*(2), 687–698. https://doi.org/10.1007/s11192-018-2981-6

Larivière, V., Haustein, S., & Mongeon, P. (2015a). Big publishers, bigger profits: How the scholarly community lost the control of its journals. *MediaTropes, 5*(2), 102–110.

Larivière, V., Haustein, S., & Mongeon, P. (2015b). The oligopoly of academic publishers in the digital era. *PLOS ONE, 10*(6), e0127502–e0127502. https://doi.org/10.1371/journal.pone.0127502

Lemke, T. (2002). Foucault, governmentality, and critique. *Rethinking Marxism, 14*(3), 49–64. https://doi.org/10.1080/089356902101242288

Levine, H. S. (1967). Economics. In G. Fischer (Ed.), *Science and ideology in soviet society 1917–1967* (pp. 107–138). Atherton Press.

Lewis, J. M. (2015). The politics and consequences of performance measurement. *Policy and Society, 34*(1), 1–12. https://doi.org/10.1016/j.polsoc.2015.03.001

Lewis, R. (1979). *Science and industrialisation in the USSR*. Palgrave Macmillan UK. https://doi.org/10.1007/978-1-349-03786-5

Leydesdorff, L., Wagner, C. S., & Bornmann, L. (2019). Diversity measurement: Steps towards the measurement of interdisciplinarity? *Journal of Informetrics.* https://doi.org/10.1016/j.joi.2019.03.016

Li, L. (2009). *Research priorities and priority-setting in China*. Vinnova. www.vinnova.se/contentassets/44b342da5cdd4c4082b51c391d4ee423/va-09-21.pdf?cb=20170915134441

Lin, J. (2020). Altmetrics gaming: Beast within or without? In M. Biagioli & A. Lippman (Eds.), *Gaming the metrics: Misconduct and manipulation in academic research* (pp. 213–227). MIT Press.

Liu, X., & Chen, X. (2018). Journal retractions: Some unique features of research misconduct in China. *Journal of Scholarly Publishing, 49*(3), 305–319. https://doi.org/10.3138/jsp.49.3.02

Liu, Y. (2012). Strategies for developing Chinese university journals through a comparison to western academic journal publishing. *Serials Review, 38*(2), 76–79. https://doi.org/10.1080/00987913.2012.10765432

López Piñeiro, C., & Hicks, D. (2015). Reception of Spanish sociology by domestic and foreign audiences differs and has consequences for evaluation. *Research Evaluation, 24*(1), 78–89. https://doi.org/10.1093/reseval/rvu030

Lovakov, A., Panova, A., Sterligov, I., & Yudkevich, M. (2021). Does government support of a few leading universities have a broader impact on the higher education system? Evaluation of the Russian university excellence initiative. *Research Evaluation, 30*(3), 240–255. https://doi.org/10.1093/reseval/rvab006

Lowe, T., & Wilson, R. (2017). Playing the game of outcomes-based performance management. Is gamesmanship inevitable? Evidence from theory and practice. *Social Policy and Administration, 51*(7), 981–1001. https://doi.org/10.1111/spol.12205

Lucas, L. (2006). *The research game in academic life*. The Society for Research into Higher Education & Open University Press.

Luczaj, K. (2020). Foreign-born scholars in Central Europe: A planned strategy or a 'dart throw'? *Journal of Higher Education Policy and Management, 6*(42), 602–616. https://doi.org/10.1080/1360080X.2019.1682955

Lukes, S. (1974). *Power: A radical view*. The Macmillan Press.

Luo, Y. (2013). Building world-class universities in China. In J. C. Shin & B. M. Kehm (Eds.), *Institutionalization of world-class university in global competition* (pp. 165–183). Springer Netherlands. https://doi.org/10.1007/978-94-007-4975-7

Macháček, V., & Srholec, M. (2017). *Predatory journals in Scopus*. Institute for democracy and economic analysis. https://idea-en.cerge-ei.cz/files/IDEA_Study_2_2017_Predatory_journals_in_Scopus/files/downloads/IDEA_Study_2_2017_Predatory_journals_in_Scopus.pdf

Marcella, R., Lockerbie, H., & Bloice, L. (2016). Beyond REF 2014: The impact of impact assessment on the future of information research. *Journal of Information Science, 42*(3), 369–385. https://doi.org/10.1177/0165551516636291

Marewski, J. N., & Bornmann, L. (2018). *Opium in science and society: Numbers*. https://doi.org/10.48550/arXiv.1804.11210

Marginson, S. (1997). Steering from a distance: Power relations in Australian higher education. *Higher Education, 34*(1), 63–80. https://doi.org/10.1023/A:1003082922199

Marginson, S. (2018). The new geo-politics of higher education: Global cooperation, national competition and social inequality in the World-Class University (WCU) sector [Working Paper, no. 34]. www.researchcghe.org/perch/resources/publications/wp34final.pdf

Marginson, S., & Rhoades, G. (2002). Beyond national states, markets, and systems of higher education: A glonacal agency heuristic. *Higher Education, 43*(3), 281–309. https://doi.org/10.1023/A:1014699605875

Mari, L., Maul, A., & Wilson, M. (2019). Can there be one meaning of "measurement" across the sciences? *Journal of Physics: Conference Series, 1379*(1), 012022. https://doi.org/10.1088/1742-6596/1379/1/012022

Marini, G. (2018). Tools of individual evaluation and prestige recognition in Spain: How sexenio "mints the golden coin of authority." *European Journal of Higher Education, 8*(2), 201–214. https://doi.org/10.1080/21568235.2018.1428649

Martin, A., & Martin, T. (2016). A not-so-harmless experiment in predatory open access publishing: An experiment in predatory open access publishing. *Learned Publishing, 29*(4), 301–305. https://doi.org/10.1002/leap.1060

Martin, B. R. (1996). The use of multiple indicators in the assessment of basic research. *Scientometrics, 36*(3), 343–362. https://doi.org/10.1007/BF02129599

Marzolla, M. (2016). Assessing evaluation procedures for individual researchers: The case of the Italian National Scientific Qualification. *Journal of Informetrics, 10*(2), 408–438. https://doi.org/10.1016/j.joi.2016.01.009

Mayer, I. S. (2009). The gaming of policy and the politics of gaming: A review. *Simulation and Gaming, 40*(6), 825–862. https://doi.org/10.1177/1046878109346456

Mazov, N. A., Gureev, V. N., & Kalenov, N. E. (2018). Some assessments of the list of journals in the Russian Science Citation Index. *Herald of the Russian Academy of Sciences, 88*(2), 133–141. https://doi.org/10.1134/S1019331618020053

McCrostie, J. (2018). Predatory conferences: A case of academic cannibalism. *International Higher Education, 2*(93), 6–8. https://doi.org/10.6017/ihe.0.93.10415

McCulloch, S. (2017, February 9). The importance of being REF-able: Academic writing under pressure from a culture of counting. *The London School of Economics and Political Science*. https://blogs.lse.ac.uk/impactofsocialsciences/2017/02/09/the-importance-of-being-ref-able-academic-writing-under-pressure-from-a-culture-of-counting/

McKiernan, E. C., Schimanski, L. A., Muñoz Nieves, C., Matthias, L., Niles, M. T., & Alperin, J. P. (2019). Use of the journal impact factor in academic review, promotion, and tenure evaluations. *eLife, 8*, e47338. https://doi.org/10.7554/eLife.47338

McLuhan, M. (1994). *Understanding media: The extensions of man*. MIT Press.

Mead, G. H. (1934). Play, the game, and the generalized other. In G. H. Mead (Ed.), *Mind, self and society: From the standpoint of a social behaviorist* (pp. 158–159). University of Chicago Press.

Meho, L. I. (2020). Highly prestigious international academic awards and their impact on university rankings. *Quantitative Science Studies, 1*(2), 824–848. https://doi.org/10.1162/qss_a_00045

Mennicken, A., & Espeland, W. N. (2019). What's new with numbers? Sociological approaches to the study of quantification. *Annual Review of Sociology, 45*(1), 223–245. https://doi.org/10.1146/annurev-soc-073117-041343

Merton, R. K. (1936). The unanticipated consequences of purposive social action. *American Sociological Review, 1*(6), 894–904. https://doi.org/10.2307/2084615

Merton, R. K. (1968). The Matthew effect in science: The reward and communication systems of science are considered. *Science, 159*(3810), 56–63. https://doi.org/10.1126/science.159.3810.56

Merton, R. K. (1973). *The sociology of science: Theoretical and empirical investigations* (N. W. Storer, Ed.). The University of Chicago Press.

Ministerstwo Nauki i Szkolnictwa Wyższego. (2018). *Ewaluacja jakości działalności naukowej: Przewodnik*. Ministerstwo Nauki i Szkolnictwa Wyższego. www.gov.pl/attachment/c28d4c75-a14e-46c5-bf41-912ea28cda5b

Mirskaya, E. Z. (1995). Russian academic science today: Its societal standing and the situation within the scientific community. *Social Studies of Science, 25*(4), 705–725. https://doi.org/10.1177/030631295025004006

Mishler, W., & Rose, R. (1997). Trust, distrust and skepticism: Popular evaluations of civil and political institutions in post-communist societies. *The Journal of Politics, 59*(2), 418–451. https://doi.org/10.1017/S0022381600053512

Moed, H. F. (2017). A critical comparative analysis of five world university rankings. *Scientometrics, 110*(2), 967–990. https://doi.org/10.1007/s11192-016-2212-y

Moher, D., Shamseer, L., Cobey, K. D., Lalu, M. M., Galipeau, J., Avey, M. T., Ahmadzai, N., Alabousi, M., Barbeau, P., Beck, A., Daniel, R., Frank, R., Ghannad, M., Hamel, C., Hersi, M., Hutton, B., Isupov, I., McGrath, T. A., McInnes, M. D. F., Page, M. J., Pratt, M., Pussegoda, K., Shea, B., Srivastava, A., Stevens, A., Thavorn, K., Van Katwyk, S., Ward, R., Wolfe, D., Yazdi, F., Yu, A. M., & Ziai, H. (2017). Stop this waste of people, animals and money. *Nature*, *549*(7670), 23–25. https://doi.org/10.1038/549023a

Moore, P. V. (2017). *The quantified self in precarity: Work, technology and what counts*. Routledge.

Moosa, I. (2018). *Publish or perish. Perceived benefits versus unintended consequences*. Edward Elgar Publishing. https://doi.org/10.4337/9781786434937

Mouritzen, P. E., & Opstrup, N. (2020). *Performance management at universities: The Danish bibliometric research indicator at work*. Springer International Publishing. https://doi.org/10.1007/978-3-030-21325-1

Moxham, N., & Fyfe, A. (2018). The royal society and the prehistory of peer review, 1665–1965. *Historical Journal*, *61*(4), 863–889. https://doi.org/10.1017/S0018246X17000334

Muller, J. Z. (2018). *The tyranny of metrics*. Princeton University Press.

Müller, M. (2018). In search of the Global East: Thinking between North and South. *Geopolitics*, *25*(3), 1–22. https://doi.org/10.1080/14650045.2018.1477757

Müller, R., & De Rijcke, S. (2017). Exploring the epistemic impacts of academic performance indicators in the life sciences. *Research Evaluation*, *26*(3), 157–168. https://doi.org/10.1093/reseval/rvx023

Münch, R. (2013). *Academic capitalism: Universities in the global struggle for excellence*. Routledge.

Musselin, C. (2014). Towards a European academic labour market? Some lessons drawn from empirical studies on academic mobility. *Higher Education*, *48*(1), 55–78.

Musselin, C. (2018). New forms of competition in higher education. *Socio-Economic Review*, *16*(3), 657–683. https://doi.org/10.1093/ser/mwy033

Nabout, J. C., Parreira, M. R., Teresa, F. B., Carneiro, F. M., Da Cunha, H. F., De Souza Ondei, L., Caramori, S. S., & Soares, T. N. (2014). Publish (in a group) or perish (alone): The trend from single- to multi-authorship in biological papers. *Scientometrics*, *102*(1), 357–364. https://doi.org/10.1007/s11192-014-1385-5

Nalimov, V. V., & Mulchenko, Z. M. (1969). *Naukometrija: Izučenie razvitija nauki kak informacionnogo processa*. Nauka.

Nature. (2020). China's research-evaluation revamp should not mean fewer international collaborations. *Nature*, *579*(7797), 8. https://doi.org/10.1038/d41586-020-00625-0

Nazarovets, S. (2020). Controversial practice of rewarding for publications in national journals. *Scientometrics*, *124*(1), 813–818. https://doi.org/10.1007/s11192-020-03485-7

Neal, H. A., Smith, T. L., & McCormick, J. B. (2008). *Beyond Sputnik: U.S. science policy in the twenty-first century*. University of Michigan Press.

Neave, G. (1998). The evaluative state reconsidered. *European Journal of Education*, *33*(3), 265–284. www.jstor.org/stable/1503583

Neave, G. (2012). *The evaluative state, institutional autonomy and re-engineering higher education in Western Europe: The prince and his pleasure*. Palgrave Macmillan.

Neff, M. W. (2018). Publication incentives undermine the utility of science: Ecological research in Mexico. *Science and Public Policy*, *45*(2), 191–201. https://doi.org/10.1093/scipol/scx054

Noga-Styron, K. E., Olivero, J. M., & Britto, S. (2017). Predatory journals in the criminal justices sciences: Getting our cite on the target. *Journal of Criminal Justice Education*, *28*(2), 174–191. https://doi.org/10.1080/10511253.2016.1195421

Nolting, L. E. (1978). *The planning of research, development, and innovation in USSR*. US Department of Commerce, Bureau of the Census.

Ochsner, M., Kulczycki, E., & Gedutis, A. (2018, September 12–14). *The diversity of European research evaluation systems* [Paper presentation]. 23rd International Conference on Science and Technology Indicators, Leiden, The Netherlands. http://repozytorium.amu.edu.pl:8080/bitstream/10593/24096/1/STI2018_paper_204.pdf

O'Connell, B. T., De Lange, P., Martin-Sardesai, A., & Agyemang, G. (2020a). Measurement and assessment of accounting research, impact and engagement. *Accounting, Auditing & Accountability Journal, 33*(6), 1177–1192. https://doi.org/10.1108/AAAJ-05-2020-4560

O'Connell, B. T., De Lange, P., Stoner, G., & Sangster, A. (2020b). Impact of research assessment exercises on research approaches and foci of accounting disciplines in Australia. *Accounting, Auditing & Accountability Journal, 33*(6), 1277–1302. https://doi.org/10.1108/AAAJ-12-2019-4293

OECD. (2015). *Frascati manual 2015: Guidelines for collecting and reporting data on research and experimental development.* OECD Publishing.

Olechnicka, A., Ploszaj, A., & Celi, D. (2019). *The geography of scientific collaboration.* Routledge.

Olssen, M., & Peters, M. A. (2005). Neoliberalism, higher education and the knowledge economy: From the free market to knowledge capitalism. *Journal of Education Policy, 20*(3), 313–345. https://doi.org/10.1080/02680930500108718

Omenn, G. S. (2006). Grand challenges and great opportunities in science, technology, and public policy. *Science, 314*(5805), 1696–1704. https://doi.org/10.1126/science.1135003

Omobowale, A. O., Akanle, O., Adeniran, A. I., & Adegboyega, K. (2014). Peripheral scholarship and the context of foreign paid publishing in Nigeria. *Current Sociology, 62*(5), 666–684. https://doi.org/10.1177/0011392113508127

Önder, Ç., & Erdil, S. E. (2017). Opportunities and opportunism: Publication outlet selection under pressure to increase research productivity. *Research Evaluation, 26*(2), 66–77. https://doi.org/10.1093/reseval/rvx006

Oppenheim, C. (2008). Out with the old and in with the new: The RAE, bibliometrics and the new REF. *Journal of Librarianship and Information Science, 40*(3), 147–149. https://doi.org/10.1177/0961000608092550

Ossowska, M., & Ossowski, S. (1935). Nauka o nauce. *Nauka Polska, 20,* 1–12.

Ossowska, M., & Ossowski, S. (1964). The science of science. *Minerva, 3*(1), 72–82. https://doi.org/10.1007/BF01630150

Osuna, C., Cruz-Castro, L., & Sanz-Menéndez, L. (2011). Overturning some assumptions about the effects of evaluation systems on publication performance. *Scientometrics, 86*(3), 575–592. https://doi.org/10.1007/s11192-010-0312-7

Pace, I. (2018, April 3). The RAE and REF: Resources and critiques. *Desiring progress.* https://ianpace.wordpress.com/2018/04/03/the-rae-and-ref-resources-and-critiques/

Pan, R. K., & Fortunato, S. (2014). Author impact factor: Tracking the dynamics of individual scientific impact. *Scientific Reports, 4*(1), 4880. https://doi.org/10.1038/srep04880

Pardo Guerra, J. P. (2020). Research metrics, labor markets, and epistemic change: Evidence from Britain 1970–2018. SocArXiv. https://doi.org/10.31235/osf.io/yzkfu

Parr, C. (2014, December 3). Imperial College professor Stefan Grimm "was given grant income target." *Times Higher Education.* www.timeshighereducation.com/news/imperial-college-professor-stefan-grimm-was-given-grant-income-target/2017369.article

Parsons, T. (1987). *On institutions and social evolution: Selected writings* (L. Mayhew, Ed.) (3rd ed.). University of Chicago Press.

Pecorari, D. (2021). Predatory conferences: What are the signs? *Journal of Academic Ethics, 19,* 343–361. https://doi.org/10.1007/s10805-021-09406-4

Perlin, M. S., Imasato, T., & Borenstein, D. (2018). Is predatory publishing a real threat? Evidence from a large database study. *Scientometrics, 116*(1), 255–273. https://doi.org/10.1007/s11192-018-2750-6

Perry, J. L., & Engbers, T. A. (2009). Back to the future? Performance-related pay, empirical research, and the perils of persistence. *Public Administration Review, 69*(1), 39–51. https://doi.org/10.1111/j.1540-6210.2008.01939_2.x

Petr, M., Engels, T. C. E., Kulczycki, E., Dušková, M., Guns, R., Sieberová, M., & Sivertsen, G. (2021). Journal article publishing in the social sciences and humanities: A comparison of Web of Science coverage for five European countries. *PLOS ONE, 16*(4), e0249879. https://doi.org/10.1371/journal.pone.0249879

Pielke, R. (2014). In retrospect: The social function of science. *Nature, 507*(7493), 427–428. https://doi.org/10.1038/507427a

Pisár, P., & Šipikal, M. (2017). Negative effects of performance based funding of universities: The case of Slovakia. *NISPAcee Journal of Public Administration and Policy, 10*(2), 171–189. https://doi.org/10.1515/nispa-2017-0017

Plane, B. (1999). The "Sputnik myth" and dissent over scientific policies under the new economic system in East Berlin, 1961–1964. *Minerva, 37*(1), 45–62.

Platonova, D., & Semyonov, D. (2018). Russia: The institutional landscape of Russian Higher education. In J. Huisman, A. Smolentseva, & I. Froumin (Eds.), *25 years of transformations of higher education systems in post-Soviet countries* (pp. 337–362). Springer International Publishing. https://doi.org/10.1007/978-3-319-52980-6_13

Polanyi, M. (1962). The republic of science: Its political and economic theory. *Minerva, 1*(1), 54–73.

Pollitt, C. (2013). The logics of performance management. *Evaluation, 19*(4), 346–363. https://doi.org/10.1177/1356389013505040

Pölönen, J. (2018). Applications of, and experiences with, the Norwegian model in Finland. *Journal of Data and Information Science, 3*(4), 31–44. https://doi.org/10.2478/jdis-2018-0019

Pölönen, J., Guns, R., Kulczycki, E., Sivertsen, G., & Engels, T. C. E. (2020). National lists of scholarly publication channels: An overview and recommendations for their construction and maintenance. *Journal of Data and Information Science, 6*(1), 50–86. https://doi.org/10.2478/jdis-2021-0004

Pölönen, J., Pylvänäinen, E., Aspara, J., Puuska, H.-M., & Rinne, R. (2021). *Publication Forum 2010–2020: Self-evaluation report of the Finnish quality classification system of peer-reviewed publication channels*. Helsinki: Federation of Finnish Learned Societies. https://doi.org/10.23847/isbn.9789525995442

Porter, T. M. (1995). *Trust in numbers: The pursuit of objectivity in science and public life*. Princeton University Press.

Power, M. (1999). *The audit society: Rituals of verification*. Oxford University Press.

Power, M. (2004). Counting, control and calculation: Reflections on measuring and management. *Human Relations, 57*(6), 765–783. https://doi.org/10.1177/0018726704044955

Price, D. J. (1963). *Little science, big science*. Columbia University Press.

Priem, J., Taraborelli, D., Groth, P., & Neylon, C. (2010, October 26). *Altmetrics: A manifesto. Altemetrics*. http://altmetrics.org/manifesto

Pusser, B., & Marginson, S. (2013). University rankings in critical perspective. *The Journal of Higher Education, 84*(4), 544–568. https://doi.org/10.1353/jhe.2013.0022

Quan, W., Chen, B., & Shu, F. (2017). Publish or impoverish: An investigation of the monetary reward system of science in China (1999–2016). *Aslib Journal of Information Management, 69*(5), 486–502. https://doi.org/10.1108/AJIM-01-2017-0014

Radosevic, S. (2003). Patterns of preservation, restructuring and survival: Science and technology policy in Russia in post-Soviet era. *Research Policy, 32*(6), 1105–1124.

Reale, E., Avramov, D., Canhial, K., Donovan, C., Flecha, R., Holm, P., Larkin, C., Lepori, B., Mosoni-Fried, J., Oliver, E., Primeri, E., Puigvert, L., Scharnhorst, A., Schubert, A.,

Soler, M., Soòs, S., Sordé, T., Travis, C., & Van Horik, R. (2017). A review of literature on evaluating the scientific, social and political impact of social sciences and humanities research. *Research Evaluation, 27*(4), 298–308. https://doi.org/10.1093/reseval/rvx025

Reitz, T. (2017). Academic hierarchies in neo-feudal capitalism: How status competition processes trust and facilitates the appropriation of knowledge. *Higher Education, 73*(6), 871–886. https://doi.org/10.1007/s10734-017-0115-3

Reuveny, R. X., & Thompson, W. R. (2007). The North-South divide and international studies: A symposium. *International Studies Review, 9*(4), 556–564. https://doi.org/10.1111/j.1468-2486.2007.00722.x

Reymert, I. (2020). Bibliometrics in academic recruitment: A screening tool rather than a game changer. *Minerva, 59*, 53–78. https://doi.org/10.1007/s11024-020-09419-0

Ridgway, V. F. (1956). Dysfunctional consequences of performance measurements. *Administrative Science Quarterly, 1*(2), 240–247.

Rochmyaningsih, D. (2019). How to shine in Indonesian science? Game the system. *Science, 363*(6423), 111–112. https://doi.org/10.1126/science.363.6423.111

Rodriguez Medina, L. (2014). *Centers and peripheries in knowledge production.* Routledge Taylor & Francis Group.

Rose, N. (1991). Governing by numbers: Figuring out democracy. *Accounting, Organizations and Society, 16*(7), 673–692. https://doi.org/10.1016/0361-3682(91)90019-B

Rosli, A., & Rossi, F. (2016). Third-mission policy goals and incentives from performance-based funding: Are they aligned? *Research Evaluation, 25*(4), 427–441. https://doi.org/10.1093/reseval/rvw012

Rousseau, S., & Rousseau, R. (2017). Being metric-wise: Heterogeneity in bibliometric knowledge. *El Profesional de La Informacion, 26*(3), 480–487. https://doi.org/10.3145/epi.2017.may.14

Rowlands, J., & Gale, T. (2019). National research assessment frameworks, publication output targets and research practices: The compliance-habitus effect. *Beijing International Review of Education, 1*(1), 138–161. https://doi.org/10.1163/25902547-00101011

Rowlands, J., & Wright, S. (2019). Hunting for points: The effects of research assessment on research practice. *Studies in Higher Education, 46*(9), 1–15. https://doi.org/10.1080/03075079.2019.1706077

Rowlands, J., & Wright, S. (2020). The role of bibliometric research assessment in a global order of epistemic injustice: A case study of humanities research in Denmark. *Critical Studies in Education*, 1–17. https://doi.org/10.1080/17508487.2020.1792523

Rozkosz, E. A. (2017). Polskie czasopisma pedagogiczne w "Wykazach czasopism punktowanych" w latach 2012, 2013 i 2015. In E. Kulczycki (Ed.), *Komunikacja naukowa w humanistyce* (pp. 153–173). Wydawnictwo Naukowe Instytutu Filozofii UAM.

Salager-Meyer, F. (2015). Peripheral scholarly journals: From locality to globality. *Ibérica, 30*(2015), 15–36.

Sandoval-Romero, V., & Larivière, V. (2019). The national system of researchers in Mexico: Implications of publication incentives for researchers in social sciences. *Scientometrics, 122*(1), 99–126. https://doi.org/10.1007/s11192-019-03285-8

Sauder, M., & Espeland, W. N. (2009). The discipline of rankings: Tight coupling and organizational change. *American Sociological Review, 74*(1), 63–82. https://doi.org/10.1177/000312240907400104

Sayer, D. (2015). *Rank hypocrisies: The insult of the REF.* Sage.

Schaff, A. (1956). Nauki Filozoficzne. In B. Suchodolski, W. Michajłow, E. Olszewski, L. Sosnowski, & S. Żółkiewski (Eds.), *Dziesięć lat rozwoju nauki w Polsce Ludowej* (pp. 87–110). Państwowe Wydawnictwo Naukowe.

Schlegel, F. (Ed.). (2015). *UNESCO science report: Towards 2030.* UNESCO Publ.

Schneider, J. W. (2009). An outline of the bibliometric indicator used for performance-based funding of research institutions in Norway. *European Political Science*, 8(3), 364–378. https://doi.org/10.1057/eps.2009.19

Schneider, J. W., Aagaard, K., & Bloch, C. W. (2016). What happens when national research funding is linked to differentiated publication counts? A comparison of the Australian and Norwegian publication-based funding models. *Research Evaluation*, 25(3), 244–256. https://doi.org/10.1093/reseval/rvv036

Scopus. (2014, November 19). Times higher education choose scopus data for its world university ranking. *Blog Scopus*. https://blog.scopus.com/posts/times-higher-education-choose-scopus-data-for-its-world-university-ranking

Serger, S. S., & Breidne, M. (2007). China's fifteen-year plan for science and technology: An assessment. *Asia Policy*, 4(1), 135–164. https://doi.org/10.1353/asp.2007.0013

Shapin, S. (2012). The ivory tower: The history of a figure of speech and its cultural uses. *The British Journal for the History of Science*, 45(1), 1–27. https://doi.org/10.1017/S0007087412000118

Sharma, Y. (2010, May 16). Asia: Governments should ignore rankings: Quacquarelli. *University World News*. www.universityworldnews.com/post.php?story=20100514204441858

Shaw, M. A. (2019). Public accountability versus academic independence: Tensions of public higher education governance in Poland. *Studies in Higher Education*, 44(12), 2235–2248. https://doi.org/10.1080/03075079.2018.1483910

Shen, C., & Björk, B.-C. (2015). "Predatory" open access: A longitudinal study of article volumes and market characteristics. *BMC Medicine*, 13(1), 230. https://doi.org/10.1186/s12916-015-0469-2

Shore, C., & Wright, S. (2003). Coercive accountability: The rise of audit culture in higher education. In M. Strathern (Ed.), *Audit cultures: Antropological studies in accountability, ethics and the academy* (pp. 69–101). Routledge.

Shu, F., Quan, W., Chen, B., Qiu, J., Sugimoto, C. R., & Larivière, V. (2020). The role of web of science publications in China's tenure system. *Scientometrics*, 122(3), 1683–1695. https://doi.org/10.1007/s11192-019-03339-x

Sīle, L., Pölönen, J., Sivertsen, G., Guns, R., Engels, T. C. E. E., Arefiev, P., Dušková, M., Faurbæk, L., Holl, A., Kulczycki, E., Macan, B., Nelhans, G., Petr, M., Pisk, M., Soós, S., Stojanovski, J., Stone, A., Šušol, J., & Teitelbaum, R. (2018). Comprehensiveness of national bibliographic databases for social sciences and humanities: Findings from a European survey. *Research Evaluation*, 27(4), 310–322. https://doi.org/10.1093/reseval/rvy016

Sivertsen, G. (2014). Scholarly publication patterns in the social sciences and humanities and their coverage in Scopus and Web of Science. In E. Noyons (Ed.), *Proceedings of the science and technology indicators conference 2014 Leiden "Context counts: Pathways to master big and little data"* (pp. 598–604). Universiteit Leiden.

Sivertsen, G. (2018a). Balanced multilingualism in science. *BiD: Textos Universitaris de Biblioteconomia i Documentació*, 40. https://doi.org/10.1344/BiD2018.40.25

Sivertsen, G. (2018b). The Norwegian model in Norway. *Journal of Data and Information Science*, 3(4), 3–19. https://doi.org/10.2478/jdis-2018-0017

Skalska-Zlat, M. (2001). Nalimov and the Polish way towards science of science. *Scientometrics*, 52(2), 211–223. https://doi.org/10.1023/A:1017911722525

Slaughter, S., & Leslie, L. L. (1997). *Academic capitalism: Politics, policies, and the entrepreneurial university*. Johns Hopkins University Press.

Smith, P. (1995). On the unintended consequences of publishing performance data in the public sector. *International Journal of Public Administration*, 18(2–3), 277–310. https://doi.org/10.1080/01900699508525011

Smolentseva, A. (2017). Where Soviet and neoliberal discourses meet: The transformation of the purposes of higher education in Soviet and post-Soviet Russia. *Higher Education*, 74(6), 1091–1108. https://doi.org/10.1007/s10734-017-0111-7

Smolentseva, A. (2019). Field of higher education research, Russia. In P. N. Teixeira & J. Shin (Eds.), *Encyclopedia of international higher education systems and institutions* (pp. 1–8). Springer Netherlands. https://doi.org/10.1007/978-94-017-9553-1_189-1

Sokal, A. (1996). A physicist experiments with cultural studies. *Lingua Franca*, 6(4), 62–64.

Sokolov, M. (2016). Can efforts to raise publication productivity in Russia cause a decline of international academic periodicals? *Higher Education in Russia and Beyond*, 7(1), 16–18.

Sokolov, M. (2020). A story of two national research evaluation systems: Towards a comparative sociology of quantification. SocArXiv. https://doi.org/10.31235/osf.io/6r85m

Sokolov, M. (2021). Can Russian research policy be called neoliberal? A study in the comparative sociology of quantifica. *Europe-Asia Studies*, 73(3), 1–21. https://doi.org/10.1080/09668136.2021.1902945

Sorokowski, P., Kulczycki, E., Sorokowska, A., & Pisanski, K. (2017). Predatory journals recruit fake editor. *Nature*, 543(7646), 481–483. https://doi.org/10.1038/543481a

Stephan, P. (2012). Perverse incentives. *Nature*, 484(7392), 29–31.

Stöckelová, T., & Vostal, F. (2017). Academic stratospheres-cum-underworlds: When highs and lows of publication cultures meet. *Aslib Journal of Information Management*, 69(5), 516–528. https://doi.org/10.1108/AJIM-01-2017-0013

Stockhammer, E., Dammerer, Q., & Kapur, S. (2021). The research excellence framework 2014, journal ratings and the marginalisation of heterodox economics. *Cambridge Journal of Economics*, 45(2), 243–269. https://doi.org/10.1093/cje/beaa054

Strathern, M. (1997). 'Improving ratings': Audit in the British University system. *European Review*, 5(3), 305–321. https://doi.org/10.1002/(SICI)1234-981X(199707)5:3<305::AID-EURO184>3.0.CO;2-4

Strielkowski, W. (2018a). Predatory publishing: What are the alternatives to Beall's list? *The American Journal of Medicine*, 131(4), 333–334. https://doi.org/10.1016/j.amjmed.2017.10.054

Strielkowski, W. (2018b). Setting new publishing standards after the Beall's list. *The International Journal of Occupational and Environmental Medicine*, 9(2), 108–110. https://doi.org/10.15171/ijoem.2018.1314

Strielkowski, W., & Gryshova, I. (2018). Academic publishing and «predatory» journals. *Science and Innovation*, 14(1), 5–12. https://doi.org/10.15407/scine14.01.005

Suber, P. (2012). *Open access*. MIT Press.

Suits, B. (1967). What is a game? *Philosophy of Science*, 34(2), 148–156. https://doi.org/10.1086/288138

Sum, N.-L., & Jessop, B. (2013). Competitiveness, the knowledge-based economy and higher education. *Journal of the Knowledge Economy*, 4(1), 24–44. https://doi.org/10.1007/s13132-012-0121-8

Swinnerton-Dyer, P., & Major, L. E. (2001, October 30). Living with RAEs. *The Guardian*. www.theguardian.com/education/2001/oct/30/researchassessmentexercise.highereducation

Szasz, T. S. (1974). *The myth of mental illness: Foundations of a theory of personal conduct*. Harper Perennial.

Teixeira da Silva, J. A., Sorooshian, S., & Al-Khatib, A. (2017). Cost-benefit assessment of congresses, meetings or symposia, and selection criteria to determine if they are predatory. *Walailak Journal of Science and Technology*, 14(4), 259–265.

Tenopir, C., & King, D. W. (2014). The growth of journals publishing. *The Future of the Academic Journal*, 159–178. https://doi.org/10.1533/9781780634647.159

Teodorescu, D., & Andrei, T. (2009). Faculty and peer influences on academic integrity: College cheating in Romania. *Higher Education*, *57*(3), 267–282. https://doi.org/10.1007/s10734-008-9143-3

Thompson, J. W. (2002). The death of the scholarly monograph in the humanities? Citation patterns in literary scholarship. *Libri*, *52*(3), 121–136. https://doi.org/10.1515/LIBR.2002.121

Tollefson, J. (2018, January 18). China declared largest source of research articles. *Nature*. https://doi.org/10.1038/d41586-018-00927-4

Tonta, Y. (2018). Does monetary support increase the number of scientific papers? An interrupted time series analysis. *Journal of Data and Information Science*, *3*(1), 19–39. https://doi.org/10.2478/jdis-2018-0002

Townsend, R. B. (2003, October 1). History and the future of scholarly publishing. *Perspectives*. www.historians.org/publications-and-directories/perspectives-on-history/october-2003/history-and-the-future-of-scholarly-publishing

Tuszko, A., & Chaskielewicz, S. (1968). *Organizowanie i kierowanie*. Państwowe Wydawnictwo Naukowe.

UHR. (2004). *A bibliometric model for performance-based budgeting of research institutions*. Norwegian Association of Higher Education Institutions.

Umut, A., & Soydal, İ. (2012). Dergi Kendine Atıfının Etkisi: Energy education science and technology Örneği. *Türk Kütüphaneciliği*, *26*(4), 699–714.

Van den Besselaar, P., Heyman, U., & Sandström, U. (2017). Perverse effects of output-based research funding? Butler's Australian case revisited. *Journal of Informetrics*, *11*(3), 905–918. https://doi.org/10.1016/j.joi.2017.05.016

Vanclay, J. K. (2011). An evaluation of the Australian Research Council's journal ranking. *Journal of Informetrics*, *5*(2), 265–274. https://doi.org/10.1016/j.joi.2010.12.001

Vavilov, S. I. (1948). *Soviet science: Thirty years*. *Marxists*. www.marxists.org/archive/vavilov/1948/30-years/x01.htm

Vera, H. (2008). Economic rationalization, money and measures: A Weberian perspective. In D. Chalcraft, F. Howell, M. Lopez Menendez, & H. Vera (Eds.), *Max Weber matters: Interweaving past and present* (pp. 135–147). Routledge.

Vernon, M. M., Balas, E. A., & Momani, S. (2018). Are university rankings useful to improve research? A systematic review. *PLOS ONE*, *13*(3), e0193762. https://doi.org/10.1371/journal.pone.0193762

Vinkler, P. (2010). *The evaluation of research by scientometric indicators*. Chandos.

Vishlenkova, E. (2018). Education management as an exact science (Russia, First Half of the Nineteenth Century). *Kwartalnik Historii Nauki i Techniki*, *63*(4), 93–114. https://doi.org/10.4467/0023589XKHNT.18.028.9519

Vishlenkova, E., & Ilina, K. A. (2013). Университетское делопроизводство как практика управления. Опыт России первой половины XIX в. *Вопросы Образования*, *1*, 232–255.

Visser, M., Van Eck, N. J., & Waltman, L. (2021). Large-scale comparison of bibliographic data sources: Web of Science, Scopus, Dimensions, Crossref and Microsoft Academic. *Quantitative Science Studies*, *2*(1), 20–41. https://doi.org/10.1162/qss_a_00112

Vitanov, N. K. (2016). *Science dynamics and research production: Indicators, indexes, statistical laws and mathematical models*. Springer.

Vostal, F. (2016). *Accelerating academia: The changing structure of academic time*. Palgrave Macmillan.

Wainwright, M. (1985, January 30). Oxford votes to refuse Thatcher degree. *The Guardian*. www.theguardian.com/politics/2010/jan/30/thatcher-honorary-degree-refused-oxford

Wallerstein, I. (2004). *World-systems analysis: An introduction*. Duke University Press.

Warczok, T., & Zarycki, T. (2016). *Gra peryferyjna: Polska politologia w globalnym polu nauk społecznych*. Wydawnictwo Naukowe Scholar.

Warren, J. (2019). How much do you have to publish to get a job in a top sociology department? Or to get tenure? Trends over a generation. *Sociological Science, 6*(7), 172–196. https://doi.org/10.15195/v6.a7

Watermeyer, R. (2019). *Competitive accountability in academic life*. Edward Elgar Publishing. https://doi.org/10.4337/9781788976138

Weber, M. (1978). *Economy and society: An outline of interpretive sociology*. University of California Press.

Wieczorek, O., & Schubert, D. (2020). The symbolic power of the research excellence framework: Evidence from a case study on the individual and collective adaptation of British Sociologists. https://doi.org/10.31235/osf.io/wda3j

Wilbers, S., & Brankovic, J. (2021). The emergence of university rankings: A historical-sociological account. *Higher Education*. https://doi.org/10.1007/s10734-021-00776-7

Williams, G., Basso, A., Galleron, I., & Lippiello, T. (2018). More, less or better: The problem of evaluating books in SSH research. In A. Bonaccorsi (Ed.), *The evaluation of research in social sciences and humanities: Lessons from the Italian experience* (pp. 133–158). Springer International Publishing.

Williams, K., & Grant, J. (2018). A comparative review of how the policy and procedures to assess research impact evolved in Australia and the UK. *Research Evaluation, 27*(2), 931–105. https://doi.org/10.1093/reseval/rvx042

Wilsdon, J., Allen, L., Belfiore, E., Campbell, P., Curry, S., Hill, S., Jones, R., Kain, R., Kerridge, S., Thelwall, M., Tinkler, J., Viney, I., Wouters, P., Hill, J., & Johnson, B. (2015). The metric tide: Report of the independent review of the role of metrics in research assessment and management. https://doi.org/10.13140/RG.2.1.4929.1363

Wilson, J. (2016). The white cube in the black box: Assessing artistic research quality in multidisciplinary academic panels. *Assessment & Evaluation in Higher Education, 41*(8), 1223–1236. https://doi.org/10.1080/02602938.2015.1075190

Wilson, M., Mari, L., & Maul, A. (2019). The status of the concept of reference object in measurement in the human sciences compared to the physical sciences. *Journal of Physics: Conference Series, 1379*(1), 012025. https://doi.org/10.1088/1742-6596/1379/1/012025

Wincewicz, A., Sulkowska, M., & Sulkowski, S. (2007). Rudolph Weigl (1883–1957): A scientist in Poland in wartime plus ratio quam vis. *Journal of Medical Biography, 15*(2), 111–115. https://doi.org/10.1258/j.jmb.2007.06-19

Woelert, P., & McKenzie, L. (2018). Follow the money? How Australian universities replicate national performance-based funding mechanisms. *Research Evaluation, 27*(3), 184–195. https://doi.org/10.1093/reseval/rvy018

Wouters, P. (1999). *The Citation Culture* [PhD thesis, University of Amsterdam]. http://hdl.handle.net/11245/1.163066

Wouters, P. (2017). Bridging the evaluation gap. *Engaging Science, Technology, and Society, 3*, 108–118. https://doi.org/10.17351/ests2017.115

Wroński, M. (2021, February). Jak zapełnić sloty. Forum Akademickie, 2. https://miesiecznik.forumakademickie.pl/czasopisma/fa-2-2021/jak-zapelnic-sloty%e2%80%a9/

Xia, J. (2017). Assessment: Which? In J. Xia (Ed.), *Scholarly communication at the crossroads in China* (pp. 139–155). Chandos Publishing. https://doi.org/10.1016/B978-0-08-100539-2.00006-8

Xia, J., Harmon, J. L., Connolly, K. G., Donnelly, R. M., Anderson, M. R., & Howard, H. A. (2015). Who publishes in "predatory" journals? *Journal of the Association for Information Science and Technology, 66*(7), 1406–1417. https://doi.org/10.1002/asi.23265

Xu, X. (2020). China "goes out" in a centre–periphery world: Incentivizing international publications in the humanities and social sciences. *Higher Education*, *80*, 157–172. https://doi.org/10.1007/s10734-019-00470-9

Xu, X., Rose, H., & Oancea, A. (2021). Incentivising international publications: Institutional policymaking in Chinese higher education. *Studies in Higher Education*, *46*(6) 1–14. https://doi.org/10.1080/03075079.2019.1672646

Yan, J. R., Baldawi, H., Lex, J. R., Simchovich, G., Baisi, L.-P., Bozzo, A., & Ghert, M. (2018). Predatory publishing in orthopaedic research: *The Journal of Bone and Joint Surgery*, 100(21), e138. https://doi.org/10.2106/JBJS.17.01569

Yang, M., & Leibold, J. (2020). Building a "Double First-class University" on China's Qing-Zang Plateau: Opportunities, strategies and challenges. *The China Quarterly*, *244*, 1140–1159. doi:10.1017/S030574102000106X

Yang, X., & You, Y. (2018). How the world-class university project affects scientific productivity? Evidence from a survey of faculty members in China. *Higher Education Policy*, *31*(4), 583–605. https://doi.org/10.1057/s41307-017-0073-5

Yudkevich, M., Altbach, P. G., & Rumbley, L. E. (Eds.). (2016). *The global academic rankings game. Changing institutional policy, practice, and academic life*. Routledge.

Zacharewicz, T., Lepori, B., Reale, E., & Jonkers, K. (2019). Performance-based research funding in EU Member States: A comparative assessment. *Science and Public Policy*, *June*, *46*(1), 105–115. https://doi.org/10.1093/scipol/scy041

Zarycki, T. (2014). *Ideologies of eastness in central and Eastern Europe*. Routledge.

Zastrow, M. (2019, November 6). South Korea clamps down on academics attending "weak" conferences. *Nature*. https://10.1038/d41586-019-03372-z

Zhang, L., & Sivertsen, G. (2020). The new research assessment reform in China and its implementation. *Scholarly Assessment Reports*, *2*(1), 3. www.scholarlyassessment reports.org/articles/10.29024/sar.15/

Znaniecki, F. (1925). Przedmiot i zadania nauki o wiedzy. *Nauka Polska*, *5*, 1–78.

Znaniecki, F. (1934). *The method of sociology*. Reinehart & Company.

Zsindely, S., Schubert, A., & Braun, T. (1982). Editorial gatekeeping patterns in international science journals: A new science indicator. *Scientometrics*, *4*(1), 57–68. https://doi .org/10.1007/BF02098006

Index